VISUALIZATION IN
GEOGRAPHICAL INFORMATION SYSTEMS

The Association for Geographic Information

The Association for Geographic Information promotes the effective use of geographic information and associated technologies. It represents the wide range of interests that make up the geographic information community and it promotes the GIS community among policy-makers. The AGI is made up of members drawn from a wide range of fields, including central and local government users, vendors, consultants and academia. It is supported by companies, organisations and interested individuals.

The aims of the AGI are:

- to influence, by developing policies for the growth of GIS and communicating these to government, the private sector and academia.
- to inform, by various means including a national conference, local meetings, and publications.
- by direct action the AGI promotes activities that will be of benefit to the broad community, such as in standards. It often acts as the focus for collective action, as in the areas of copyright or Ordnance Survey charging policy.

Activities of the AGI include:

A highly regarded Annual Conference which provides a unique forum for those who wish to hear the latest developments in UK and European policy, review research and development during the past year, and discuss applications and requirements with the leading companies in the GIS arena.

Production of a wide range of publications on various aspects of GIS. The AGI also produces a Yearbook and a quarterly Newsletter for members.

The AGI sponsors seminars and workshops held throughout the country. Recently we sponsored a series of seminars run by ESRC.

A series of regional meetings in order to reach members and non-members alike and give them the opportunity to discuss issues relevant to their own experiences within the GIS field.

An annual award for Technological Progress, Student of the Year and Journalist of the Year.

A forum for the discussion of new ideas and dissemination of information.

Active lobbying which promotes the geographic information community in a wider context.

If you need more information on the AGI please contact: 12 Great George Street, London SW1P 3AD Telephone 071 222 7000 x 472

VISUALIZATION IN GEOGRAPHICAL INFORMATION SYSTEMS

edited by

HILARY M. HEARNSHAW & DAVID J. UNWIN

JOHN WILEY & SONS

CHICHESTER · NEW YORK · BRISBANE · TORONTO · SINGAPORE

Copyright © The Association for Geographic Information 1994

Published by John Wiley & Sons Ltd,
         Baffins Lane, Chichester,
         West Sussex PO19 1UD, England
Telephone   National Chichester (0243) 779777
         International +44 243 779777

Reprinted October 1994

Other Wiley Editorial Offices

John Wiley & Sons, Inc., 605 Third Avenue,
New York, NY 10158-0012, USA

Jacaranda Wiley Ltd, 33 Park Road, Milton,
Queensland 4064, Australia

John Wiley & Sons (Canada) Ltd, 22 Worcester Road,
Rexdale, Ontario M9W 1L1, Canada

John Wiley & Sons (SEA) Pte Ltd, 37 Jalan Pemimpin #05-04,
Block B, Union Industrial Building, Singapore 2057

Library of Congress Cataloging-in-Publication Data

A CIP catalog record for this book is available from the Library of Congress

British Library Cataloguing in Publication Data

A catalogue record for this book is available from the British Library

ISBN 0-471-94435-1

Typeset in Sabon 10/11pt by Saxon Graphics Ltd, Derby
Printed and bound in Great Britain by Bookcraft (Bath) Ltd, Avon

Contents

v

LIST OF CONTRIBUTORS

KATE BEARD is Assistant Professor in the Department of Surveying Engineering at the University of Maine where she is research faculty with the National Center for Geographic Information and Analysis (NCGIA). She is currently working on multiple representations of spatial data and the visualization of data quality.

Department of Surveying Engineering, University of Maine-Orona, Maine, USA

IAN BISHOP works at the Centre for GIS and Modelling in Melbourne, Australia where he has developed methods for presenting life-like visualizations of potential projects as an aid in the planning process.

Centre for Geographic Information Systems and Modelling, The University of Melbourne, Parkville, Victoria 3052, Australia

IAN BRACKEN is Senior Lecturer in the Department of City and Regional Planning, University of Wales Cardiff. He is Director of the Wales and South West Regional Research Laboratory based in the Department, and an ESRC-supported training programme for users of the 1991 Census of Population. He has researched and written extensively on the socioeconomic applications of GIS.

Department of City and Regional Planning, University of Wales, Cardiff, PO Box 906, Cardiff CF1 3YN, Wales

KEN BRODLIE is Senior Lecturer in the School of Computer Studies at the University of Leeds and has had a long involvement with international standards for computer graphics. A special interest has been the validation of implementations of graphics standards, where he has contributed to the European-wide testing services for GKS and CGI implementations. His interest in scientific visualization began with an involvement in the NAG Graphics library, and has extended to research into new visualization systems that integrate numerical computation and graphics. He was the founding chairman of the UK Chapter of Eurographics and the UK Visualization Community Club, and is on the editorial board of Computer Graphics Forum.

School of Computer Studies, University of Leeds, Leeds LS2 9JT, UK

'BABS' BUTTENFIELD is Associate Professor of Geography at SUNY-Buffalo and research scientist with the NCGIA. She has also served as Director of the Geographic Information and Analysis Laboratory. Her research interests are in map generalisation and the role of expert systems in GIS and spatial analysis. Recently, she has focused upon the representation and integration of data quality information into GIS displays with a view to improving their utility for analysis and decision-making.

National Center for Geographic Information and Analysis, Department of Geography, State University New York-Buffalo, Buffalo, NY 14261, USA

LIN CHIH-CHANG is a graduate student in the Department of Electrical and Computer Engineering, University of California, Santa Barbara, and a research assistant at NCGIA.

National Center for Geographic Information and Analysis, University of California, Santa Barbara, CA 93106, USA

ANNA CROSS was Co-ordinator for the Centre for Research on Crime, Policing and the Community, University of Newcastle upon Tyne. She recently completed a PhD which involved the building and evaluation of a GIS to study childhood cancer. This work has been extended to investigate the use of GIS in a policing environment. She now works as a civil servant in Cheltenham, UK.

Centre for Research on Crime, Policing and the Community, Claremont Bridge, The University, Newcastle upon Tyne, NE1 7RU, UK

CLARE DAVIES is working with the Midlands Regional Research Laboratory at Loughborough University on aspects of human computer interface design and usability as applied to GIS systems.

ESRC Midlands Regional Research Laboratory, Department of Geography, University of Technology, Loughborough, Leicester, LE11 3TU, UK

DANIEL DORLING completed his PhD entitled The Visualization of Spatial Social Structure in 1991 at the University of Newcastle upon Tyne. He is currently a Research Associate, funded by a Joseph Rowntree Foundation Fellowship.

Department of Geography, University of Newcastle upon Tyne, Newcastle NE1 7RU, UK

JASON DYKES is a Research Assistant at the ESRC Midlands Regional Research Laboratory. Jason has a first degree in Geography from the University of Oxford and an MSc in GIS from the University of Leicester. He is currently working on a PhD on the visualization of enumerated geographic data and is a member of the Project ASSIST (Academic Support for Spatial Information Systems) team.

ESRC Midlands Regional Research Laboratory, Department of Geography, University of Leicester, Leicester LE1 7RH, UK

PETER FISHER is currently lecturer in GIS at Leicester University, having taught at Kingston Polytechnic, UK, and Kent State University, USA. His main areas of research interest are in the modelling, propagation and visualization of error in spatial data, and in Artificial Intelligence applications in GIS.

ESRC Midlands Regional Research Laboratory, Department of Geography, University of Leicester, Leicester LE1 7RH, UK

JULIAN GALLOP leads the Visualization Group in the Rutherford Appleton Laboratory Informatics Department, responsible for visualization, computer graphics and user interface software supporting UK academic engineering researchers. Julian participated in the definition of the ISO Graphics Standards for GKS and

PHIGS PLUS and their binding to programming languages. He is a co-editor, with Brodlie, of a major text on ViSC (Brodlie et al., 1992)

Informatics Department, Rutherford Appleton Laboratory, Chilton, Didcot, Oxfordshire OX11 OQX, UK

TONY GATRELL is Senior Lecturer in the Department of Geography, Lancaster University and Co-Director of the ESRC North West Regional Research Laboratory. He is interested in the applications of GIS techniques in environmental and geographical epidemiology.

North West Regional Research Laboratory, Department of Geography, University of Lancaster, Lancaster LA1 4YB, UK

MICHAEL GOODCHILD is Director, National Center for Geographic Information and Analysis, and Professor of Geography, University of California, Santa Barbara. His research interests include issues of accuracy and error in spatial data, and geographical data models.

National Center for Geographic Information and Analysis, University of California, Santa Barbara, CA93106, USA

HILARY HEARNSHAW has a background in mathematics, ergonomics and psychology and, while at the Midlands Regional Research Laboratory, University of Leicester, her work included investigating the psychological aspects of using GIS, especially users' mental models and the usability of GIS. She is now lecturing in Medical Audit at the University of Leicester Medical School.

Eli Lilly National Medical Audit Centre, Department of General Practice, Leicester General Hospital, Leicester LE5 4PW, UK

YEE LEUNG is Professor of Geography, Chinese University of Hong Kong. His interests include fuzzy reasoning and artificial intelligence applied to spatial problems, and he is author of Spatial Analysis and Planning under Imprecision (New York: North Holland, 1988).

Department of Geography, Chinese University of Hong Kong, Shatin, New Territories, Hong Kong

ALAN M. MACEACHREN is Professor of Geography at the Pennyslvania State University. His current research is directed to cognitive and semiotic questions concerning how maps work and cartographic contributions to scientific visualization including map animation.

Department of Geography, Pennsylvania State University, 302 Walker Building, University Park, PA 16802, USA

DAVID MEDYCKYJ-SCOTT is a research scientist and deputy director of the GENIE project at the Midlands Regional Research Laboratory, Loughborough University of Technology, UK. Current research interests include human–computer interfaces in GIS, organisational issues of GIS use and the design of spatial metadata systems.

ESRC Midlands Regional Research Laboratory, Department of Geography, University of Technology, Loughborough, Leicester, LE11 3TU, UK

STAN OPENSHAW has recently been appointed Professor of Human Geography in the School of Geography, Leeds University. He specialises in spatial and statistical analysis techniques, with particular interests in epidemiological, census and police-related data.

Department of Geography, The University, Leeds, LS2 9JT, UK

JIM PETCH is a geomorphologist and co-author of Physical Geography: Its Nature and Methods (Harper & Row, London, 1986) who works at Salford University. Currently he is involved with a recently established distance learning Masters course in GIS.

Department of Geography, University of Salford, Salford M5 4WT, UK

ANDREW TURK has degrees in surveying, cartography and psychology. He has worked in an Australian government mapping agency, conducted research into tactual graphics and lectured in cartography and GIS at the University of Melbourne. Andrew was recently awarded a PhD in human factors aspects of GIS.

Information Systems, School of Mathematical and Physical Sciences, Murdoch University, Perth, Western Australia 6150

DAVID UNWIN has recently been appointed Professor of Geography at Birkbeck College London. Formerly he was Senior Lecturer in the Department of Geography, University of Leicester where he was involved with the ESRC Midlands Regional Research Laboratory and the Computers in Teaching Initiative Centre for Geography. Nationally he is on the council of the Association for Geographic Information and has a long-standing interest in spatial statistical analysis and mapping.

Department of Geography, Birkbeck College, University of London, Gresse Street, London W1P 1PA, UK

MAHES VISVALINGAM is Senior Lecturer in Computer Science at the University of Hull where she coordinates the Cartographic Information Systems Research Group. She is currently Chair of the Research Committee of the British Cartographic Society.

Department of Computer Science, The University of Hull, Hull HU6 7RX, UK

DAVID WAUGH is a Research Associate at the University of Newcastle upon Tyne specialising in the development of crime pattern analysis techniques and systems for operational use in a policing environment. He is also involved in the development of computer movies using police incident log data.

Centre for Research on Crime, Policing and the Community, Claremont Bridge, The University, Newcastle upon Tyne, NE1 7RU, UK

JOSEPH WOOD is a Lecturer in Geography at the University of Leicester. His research interests include digital elevation modelling, GIS visualization techniques and the production of computer based learning materials.

ESRC Midlands Regional Research Laboratory, Department of Geography, University of Leicester, Leicester LE1 7RH, UK

MIKE WOOD is a Senior Lecturer in Geography (Cartography) in the University of Aberdeen. His research studies include map-reading and the creation and application of 3D landscape visualizations. He is a Vice-President of the International Cartographic Association and President of the Society of Cartographers.

Department of Geography, The University, Old Aberdeen, AB9 2UF, UK

INTRODUCTION: THE PROCESS

A Geographical Information System (GIS) can be defined as a system for capturing, storing, checking, manipulating, analysing and displaying data which are spatially referenced (DoE, 1987, p. 132; see also Maguire, 1991). Since its appearance sometime during the 1960s the field of computer GIS has grown rapidly, creating an enormous literature explosion in its wake. The most comprehensive summary is a recent, two-volume compendium (Maguire, Goodchild and Rhind, 1991). Although visualization has been the cornerstone of scientific progress throughout history, the term scientific visualization (ViSC) has recently come to refer to the exploration of data and information in such a way as to gain understanding and insight into these data. Generally, this is conducted in powerful three-dimensional graphical computing environments. A recent, accessible summary is provided by Earnshaw and Wiseman (1992). In the field of geography and visualization, a number of contributions have started to appear in the academic literature of which the most accessible is Monmonier and MacEachren (1992).

It is clear that there are many areas of interaction between GIS and ViSC. The basic paradigm of ViSC, that 'seeing' is a good way towards understanding, finds support in the use of maps and other graphical displays as outputs from GIS systems to isolate and describe patterns in spatial distribution. Both fields draw heavily on developments in database management, computer graphics, user interface design, image processing and so on. This book is the result of a workshop which attempted to explore this common ground under the auspices of the UK Association for Geographical Information (AGI), at Burleigh Court, University of Loughborough, UK in the summer of 1992. The workshop was unusual in that, right from the outset, its declared aim was to produce a coherent statement of the role and position of visualization, incorporating views from the worlds of computer science and ViSC, traditional cartography and GIS.

In order to understand how this book arose, it is necessary to examine the process by which it came about and the ways by which we have tried to avoid producing yet another set of only vaguely linked conference papers.

1 Every contributor was asked to produce a draft of their input, to an agreed topic, several months before the meeting took place.
2 These papers were pre-circulated to all other participants, for comments and suggestions, and were then presented in summary form at a series of initial plenary sessions.
3 The participants then split into sections and spent the next two days working on revisions to the original papers and the draft of an introduction to each planned book section, before reporting back to final plenary sessions.
4 All contributions were further revised, after the workshop, before being submitted to the editors.

The result is the book which you now see. Section A sets the scene by exploring the ground between traditional mapping concerns and ViSC. Visualization specialists will find the chapters by Wood and by Visvalingam useful introductions to the long-standing concern of cartographers for what is now termed visualization. Cartographers and GIS specialists will find that Brodlie's account locates them in the wider field of ViSC and Gallop gives them an overview of what is currently possible in present ViSC systems. Section B examines some of the ways by which computer technology has helped improve the visualization of spatial data by implementing techniques such as density estimation (Gatrell), surface modelling (Bracken), spatial autocorrelation measurement (Dykes), cartograms (Dorling), visual realism (Bishop) and animation (MacEachren; Openshaw, Waugh and Cross). After reading this section, it should be abundantly clear that ViSC is already helping spatial analysts to get much clearer and more honest maps of their data. Good visualization only makes sense if the data themselves are of reasonable quality, defined in the modern way as 'fitness for purpose'. Section C examines how data quality itself might be visualized and we make no apologies for devoting space to what Buttenfield and Beard show to be a critical aspect of GIS use. Carefully constructed displays clearly have a part to play in showing data quality (Goodchild, Chih-Chang and Leung; Fisher) and the uncertainties introduced by the algorithms used (Wood). Finally, Section D explores the human factors implicit in any visualization. In it, we explore the psychology of how we 'read' computer displays (Hearnshaw), how we interact with complex software systems (Medyckyj-Scott), and how all this relates to our scientific endeavours (Petch).

As editors, our role has been to collate the materials, checking for consistency and coherence. We hope we have produced neither a set of bound conference papers nor a multi-authored textbook, but something which lies somewhere between the two. It follows from the nature of this production process that attribution of the individual chapters is difficult. In each case we have given the name(s) of the author of the initial and final drafts but recognise that there have been inputs into each, and especially the initial section introductions, from almost everyone who attended the workshop. As the editors whose names appear on the front cover, we are, of course, ultimately responsible for any errors and omissions.

Hilary Hearnshaw and David Unwin

ACKNOWLEDGEMENTS

The workshop which gave rise to this book would not have taken place without the generous financial and moral support of the UK Association for Geographic Information through its Education, Training and Research Committee. The staff of Burleigh Court, University of Technology, Loughborough did much to help us create a superb environment in which to work. Other support was provided by the ESRC Midlands Regional Research Laboratory and Department of Geography, University of Leicester. Finally, all the participants are grateful to Dr Iain Stevenson of Belhaven Press (John Wiley), not only for his presence – as a geographer – at the workshop, but also for his enthusiastic support and advice.

REFERENCES

Department of the Environment (DoE) (1987) *Handling Geographic Information*, HMSO, London

Earnshaw R A, Wiseman N (eds) (1992) *An Introductory Guide to Scientific Visualization*, Springer-Verlag, Berlin

Maguire D J (1991) An overview and definition of GIS. In: Maguire D J, Goodchild M F, Rhind D W (eds) *Geographical Information Systems*, Longman, Harlow, pp. 2–20

Maguire D J, Goodchild M F, Rhind D W (eds) (1991) *Geographical Information Systems*, Longman, Harlow

Monmonier M, MacEachren A M (eds) (1992) Special content: Geographic Visualization. *Cartography and Geographic Information Systems* 19(4): 197–260

SECTION A
VISUALIZATION IN GIS

1 ViSC AND GIS: SOME FUNDAMENTAL CONSIDERATIONS

M. WOOD & K. BRODLIE

... BACKGROUND

The theme of this volume highlights a growing awareness within the GIS community of the development of Visualization in Scientific Computing (ViSC). A sensitivity to the possible applications of investigative techniques used in other scientific fields is not new to disciplines such as geography. With its fuzzy boundaries and overarching involvement with the description, analysis and explanation of earth-related patterns and processes, both physical and human, geography has been a notorious magpie, acquiring potentially useful ideas and techniques from many sources to add to its box of tools. A fundamental concern with two- and three-dimensional spatial relationships inevitably leads to overlaps with subjects such as geology, climatology and oceanography, not to mention the social sciences. Maps and mapping (at whatever scale) are also unifying factors. Although manual systems for the processing and display of geographical data have been in use for decades, if not centuries, computer GIS offers a completely new working environment for both geographical research and the management of the environment. Although GIS offers much more, an important analytical as well as presentational feature is the facility to generate graphics (including maps) customised for content, scale and appearance.

In science and engineering, the importance of graphic representations of data can be traced from 'scientific data plotting' to the realms of what is now known as Visualization in Scientific Computing (ViSC). This subject, even younger than computer GIS, is closely associated with faster processing hardware and more sophisticated computer graphics, including animation.

Today successful new scientific developments are frequently shared. This volume examines the interface of state-of-the-art ViSC, with its often unique terminology, focus, structure and functionality, and GIS, which has developed into a wider environment of applications from data management to scientific research. The perceived goal is to identify aspects of ViSC which might enhance the functional scope and efficiency of GIS to satisfy the expanding array of tasks to which it is being applied.

... VISUALIZATION AND ViSC

One of the paradoxes of science is that the language or jargon of its subfields can often be barriers to, rather than means of, communication. The word 'visualization' itself is a case in point. Various definitions exist, including the concept of a 'visual' as a concrete representation (a produced perceptible display). While there is no consensus as yet, visualization as a mental process seems to be commonly accepted. From this meaning we can derive a contemporary definition for the word in the context of ViSC: a set of tools (today, mainly software) used to permit visual data analysis. Thus, through images displayed on computer screens, assistance is provided to human information processing, enhancing mental visualization and the comprehension of 2D and 3D spatial relationships and spatial problems. The aim of this process is to stimulate the acquisition of insights into and solutions to the problems being addressed. These tools offer much more than mere static displays and include animation (e.g. fly-throughs of 3D space) and interaction with the data. A feedback loop exists to permit almost instantaneous responses to queries made of the screen display, for example. The data structures and linked graphics systems, permitting the efficient operation of such feedback loops, are the keys to the success of ViSC tools. 'The changing pace of interaction can result in qualitative as well as quantitative increases in productivity' (Palmer, 1992). A kind of interaction could be identified in pre-computer research methods but, through the slowness of data manipulation and graphics output, it often fell rapidly behind the speed of the mind as it explored beyond such static displays. New facilities for probing, tracking, steering and post-processing offer important advances in investigative research.

The processing stages within research have been identified by MacEachren et al. (1992) as visual exploration and confirmation (identified as visual thinking) and synthesis and presentation (classified as visual communication). A 'brainstorming' environment such as offered by ViSC must be stimulating to creative scientific enquiry. If some aspects of creative or inventive thinking (believed to lie beyond direct mental control) are highly visuo-spatial in character (the incubation/exploration stage of the research process) it is logical to enrich this phase with a variety of visual representations of data which can be explored. It is also possible to increase the investigator's sensitivity when operating within such virtual environments through the use of non-visual computer output. Spatially referenced audio, for instance, can enrich mental visualization by means of sounds which change as the operator moves a screen probe to different locations within the 3D space. This reinforcement of visual feedback could even include audio triggers to indicate passage beyond certain critical thresholds.

... HISTORICAL PERSPECTIVE AND CONTEMPORARY REVIEW OF ViSC AND GIS

Scientific visualization has been a major topic in computing over the past five years. Its origins are often traced to the publication of the much-cited National Science Foundation report on *Visualization in Scientific Computing* (McCormick et al., 1987). This argued for investment in visualization if full advantage was to be taken of the installation of supercomputers at a number of centres in the United States. Though it did not apparently lead to direct funding for visualization, several supercomputer centres initiated major research and development projects into the subject. The recent survey volume by Neilson et al. (1990) provides a good overview of these activities. In addition a number of computer manufacturers have developed powerful visualization systems, such as AVS and Iris Explorer, and these are now in use worldwide.

It is therefore tempting to think of visualization as a young subject, but in truth its origins go far back in history. It is often argued that many of the original thinkers in science did much of their thinking visually and may have had poor verbal skills. A notable early visualizer in science was James Clerk Maxwell, one of the founders of thermodynamics, who built 3D clay models to help understand the behaviour of a function of two variables, just as we do today with surface plots in visualization systems.

In the early days of computing, scientists made substantial use of computer graphics to display the results of their calculations. Pioneering work was done in the UK in the 1960s, in particular at the Rutherford and Culham Laboratories, where scientists produced animated film sequences of their data. This led to the development of scientific graphics libraries such as GHOST from Culham Laboratories which was widely used in the UK scientific research community in the 1970s. Other examples, also from the UK, were GROATS from Rutherford Laboratory and the GINO libraries from the CAD Centre in Cambridge.

These were Fortran subroutine libraries; scientists incorporated the routines within their own computational programs. In the 1980s, there was a trend away from 'writing programs' and towards the use of packages. The mode of working was for the scientist to create a dataset from the application, and pass the dataset to a menu-driven package which would offer a variety of display options.

This style of working was much simpler than writing programs, but of course it was achieved at the expense of flexibility. The next stage of evolution, indeed where we are now, is the application builder system such as AVS and Iris Explorer. Here the system provides a set of modules very similar to the subroutines provided in the old GHOST and GINO libraries, but now the programming is done in a very much simpler way. Using a 'point-and-click' interface, the scientist selects modules from the library, defines a path for data to flow between modules, and then fires the process by providing data. Scientists can include their own applications as modules, thus performing the entire computation within the visualization system.

Thus it can be argued that the new visualization systems offer merely a simpler programming model than the traditional graphics libraries. This model has the great advantage, however, that changes to the 'visual program' can be easily made, allowing the scientist to experiment with different techniques. Moreover the hardware on which these systems run is now becoming powerful enough for this 'human in the loop' interaction style to be viable.

Whereas 'true' ViSC can be viewed as a very recent development, the origins of computer GIS are slightly older. In the early stages two trends can be observed, one of which, computer-assisted cartography (CAC) had, at the time, nothing to do with GIS as defined today. CAC, related to computer-aided drafting or design (CAD), was primarily concerned with the automation of

5

mapping processes which had been practised for decades or more. Naturally a major concern was the quality of the graphic output and it might be argued that the timing of decisions to proceed with such automated systems was largely determined by the state of the art in computer graphic output. It was not until the 1960s that CRT became part of the system hardware and at that time the only commonly available graphic hardcopy output was the raster format line printer. Conventional maps were conceived as vector images and considerable developments in hardware and software engineering would have to take place before output satisfied the conventional cartographer and map user. Equally, cartographers, especially in national mapping organisations, were preoccupied by the need to convert the massive library of existing analogue maps to digital form.

During the same period another development was taking place. Analysts, including planners and landscape architects, became interested in employing the computer to aid their analytical work. They were dealing with geographical space but were more concerned with the computer's calculating powers than the quality of the output. Mapping programs such as SYMAP (Liebenberg, 1976) and IMGRID (Sinton, 1977) provided some of the earliest solutions for these scientists. Their approach also drew attention to the importance of the data, its storage structure and manipulation. Contemporary GIS has essentially combined the database management system concept with the procedures of CAC and CAD.

Lack of appreciation of the importance of flexibly structured data created delays in the progress of some digital mapping programmes. Data structures were primarily aimed at recreating the digitised map and could not respond to questions about the elements of the database. Modern GIS, exemplified by ARC/INFO, might be described as offering the best of both worlds, an integrated database management and graphics system. The more recent evolution of GIS can be compared with ViSC systems in that users, previously forced to carry out their own programming, were provided with packages of software modules and, more recently, graphic user interfaces, to extend the character and quantity of the user base. Another major convergence has been with the digital image processing of satellite remote sensing data. It is now recognised that these raster images are providing the largest single source of digital spatial data today and GIS technology has been adapted accordingly. In comparison with ViSC, GIS has acquired a greater and more rapidly growing user base. A huge market has developed for its management applications, especially in local authorities, the utilities and emergency services, and this has encouraged the improvement of user-friendly interfaces and the quality of graphic output.

The reader, however, should be in no doubt of the differences between the two types of system. Although Hall (1992) has argued that the mapping process need not be limited by the size of the object to be represented, GIS is primarily concerned with geographical, rather than molecular or cosmic, topics and scales. Its graphic output is also dominated by cartographic images of point, line, area and surface features. ViSC, on the other hand, is characterised by the representation of 3D space and is associated with much wider fields of scientific investigation, ranging from genetic engineering to oil exploration. Its methods are also being employed in contrasting areas such as clothing design and motion picture animation.

The process of evaluating and enhancing the quality of output graphics from both systems should make maximum use of all contemporary knowledge of design rules, etc., whatever their origin. Cartography provides a rich source. The significance of good graphics, even in exploratory (rather than communicative) tools, was revealed by the enthusiastic reactions of one group of British Petroleum geophysicists who, following the arrival of a new colour printer, found their previously black-and-white seismic 'contour' plots much easier to analyse with the simple application of suitably graduated coloured layer tints.

... SCIENTIFIC VISUALIZATION THROUGH GIS

It must be accepted that a GIS should have available all kinds of perceptible displays to aid geographical scientific visualization. Everything from full coloured photographs to highly diagrammatic images (and not just maps) should comprise the arsenal of such systems. Depictions of surface representations of three dimensions are growing rapidly in popularity. The availability of different depictions and different perspective views of the same dataset is assumed to be essential. Although many maps have been designed to depict the third dimension of undulating surfaces, truly 3D GIS (ie. going beyond the depiction of single surfaces) are still at the development stage. The design of visualizations is a critical aspect of GIS, especially when these images are to be employed as spatial decision support systems (SDSS). There have been significant increases in available functionality and the range of topics represented. Usage of GIS is becoming more interactive and potentially includes computer-supported cooperative work (CSCW). Hence optimisation of the design of visualization sequences is increasingly complex. Task definition and analysis procedures may be used to determine the users' cognitive processing requirements. Formal human–computer interaction (HCI) models and task/visualization taxonomies can also make a significant contribution to increasing the sophistication and rationality of visualization design based on cognitive work requirements.

Maximum levels of interaction, a feature of ViSC, are therefore recommended, preferably by data manipulation through the image on the screen. The growing real-world requirements of GIS are being constantly reviewed to identify which new facilities should be developed. The changes will be brought about by expansion of existing software, new modules or even links with existing visualization systems whose superior powers of rendering can be employed.

... THE CHAPTERS IN THIS SECTION

This introductory section includes chapters by authors from a wide variety of backgrounds and interests, reflecting many individual views and inevitably covering some similar ground. Occasional redundancies have been retained to preserve the structure of the chapters.

Chapter 2, by M. Wood, presents the case for the traditional paper map as a visualization tool. There is a tendency to assume that such a static product depicts what is known and is thus merely 'presentation' and not high on the scale of analytical usefulness. While true in a sense, practice shows that information-rich topographic maps, for instance, can provide a productive environment for further investigation. Study of these assemblies of simultaneous graphic 'overlays' of a wide range of spatial datasets can reveal correlations and lead to insights into the reasons for spatial locations and distributions. Chapter 3, by Visvalingam, examines the GIS viewpoint and makes a critical examination of the word 'visualization' in its different contexts. The use of graphics by contemporary GIS and ViSC are compared as are their potential for problem-solving and ideation. The contribution by Turk (Chapter 4) makes an in-depth analysis of GIS visualizations. The selection of suitable visualizations is viewed in the context of cognitive task analysis and taxonomies of visualization designs are reviewed. Gallop (Chapter 6) makes a structured review of current visualization software with questions posed about its applicability to GIS, and Brodlie (Chapter 5) offers an examination, classification and notation for models and visualization techniques.

... REFERENCES

Hall S S (1992) *Mapping the Next Millenium: The Discovery of New Geographies*, Random House, New York

Liebenberg E (1976) SYMAP: its uses and abuses. *Cartographic Journal* 13(1): 36–46

MacEachren A M, Butterfield B, Campbell J, DiBiase D, Monmonier M (1992) Visualization. In: Abler R, Marcus M, Olson J (eds) *Geography's Inner Worlds: Pervasive Themes in Contemporary American Geography*, Rutgers University Press, New Brunswick, New Jersey, pp. 99–137

McCormick B H, DeFanti M D, Brown M D (eds) (1987) Visualization in scientific computing. Special issue ACM SIGGRAPH *Computer Graphics* 21(6)

Neilson G M, Shriver B, Rosenblum L (eds) (1990) *Visualization in Scientific Computing*, IEEE Computer Society Press, Los Alamitos, Co., Washington

Palmer T C (1992) A language for molecular visualization. *IEEE Computer Graphics Applications* 12(3): 23–32

Sinton D F (1977) *The Users Guide to IMGRID, An Information Manipulation System for Grid Cell Structures*, Harvard University, Department of Landscape Architecture, Cambridge, Mass.

2 THE TRADITIONAL MAP AS A VISUALIZATION TECHNIQUE

M. WOOD

M. WOOD

... INTRODUCTION

Many readers will regard this chapter as a statement of the obvious. Others, perhaps including scientists active in the expanding field of ViSC, will be puzzled to think of traditional paper maps (especially pre-computer maps) as anything other than data presentation.

While specific visualization techniques adopted by individual scientists today vary, it is the author's belief that in the new age of computer graphics the role of cartography may become undervalued. It is possible that many still associate the word cartography with old craft skills rather than with the fundamental processes of mapping. If the physical dimensions of the represented object are ignored as a delimiting factor in the definition, the analysis and interpretation of the spatial patterns on, for example, a schematic representation of an x-ray photograph have much in common with typical map studies. Only the subject matter is different. The importance of mapping 'whereness' at any scale is convincingly presented by Hall (1992). Confusion arises when cartography is identified as an application of scientific visualization (often restricted to geography). In fact, it is a fundamental visualization technique, or group of techniques, in its own right, used extensively in many subject areas. Climatology, geology, archaeology and oceanography as well as fields such as chemistry (Flynn, 1990) and medicine (John et al., 1988) employ cartography to varying degrees. Nor is the technique limited solely to planimetric depiction. The 'official' definition includes cross-sections, 3D representations, models and globes. In scientific visualization terminology, cartography's powers range from presentation of the known, through analysis, to discovery of the yet unknown. Providing insights into spatially related characteristics and numbers is one of cartography's primary functions. With its long pedigree, cartography may be interpreted as having not only predated modern computer ViSC but also having originated many of its graphic methods.

... DEFINING TERMS AND PROCESSES

Haber and McNabb (1990) define Scientific Visualization (ViSC) as 'complex generalised mappings of data from raw data to rendered image'. Similarly, Palmer (1992) sees display as the penultimate stage, prior to analysis. However, graphic presentation alone is regarded as subordinate within current ViSC systems. True 'insight into numbers' (Hamming, 1962) is said to involve the whole investigative process including direct (preferably immediate) interactive access to, and manipulation of, data (Palmer, 1992). This dynamic facility is believed to offer a 'brainstorming' environment 'which reveals anomalies or unexpected trends and patterns' (Brodlie et al., 1992) during the preparatory stages of scientific enquiry. However, in the flood of excitement associated with these new possibilities, the value of pre-ViSC techniques such as cartography can easily become diminished. Indeed novice users of the new graphic tools may be unaware of how the techniques fit within the cycle of problem-solving and creative thinking.

Human cognition is essentially spatial in character and making things 'visible in the mind' has been a key to many, if not all, major scientific discoveries. Most creative thinking involves internal mental imagery and both common experience and psychological research now support the notion that not all thinking is fully accessible to direct mental control. The activities of what has been termed the 'right-brained' mode (visual, spatial, relational) are essentially hidden to the thinker whose normal open 'mental working space' is believed to be 'left-mode' (linear, logical, language/symbol-based). A model of creative thinking which embodies this 'mode' idea is offered by Edwards (1987):

1 First insight (right-mode). This is the exploratory stage involving searching out productive questions from intuitive leaps of awareness based on wide knowledge of a subject. It can relate to existing problems or involve problem-finding.
2 Saturation (left-mode). This forms the first stage of active research involving gathering, sorting and categorising information with a particular end in view.
3 Incubation (right-mode). When lines of investigation come to an end, logical analysis fails and frustration often results. The problem under consideration may seem to be set aside. The processes which continue, sometimes referred to as 'mulling-over', are unknown but may be controlled by heuristics and visual imagery. Shepard (1978) has referred to the latter as mainly entencephalic images such as memory, imagination, 'sleep-edge', dream and hallucinatory. This type of image is both complex and meaningful and seems to arise spontaneously within the brain. Edwards believes that, in drawing at least, there are non-syntactical rules and heuristics which operate and, indeed, which can be learned. The anxiety often associated with this stage may represent a left-mode reaction. The thinker, while aware that something might be happening, cannot participate and thus continues with other cognitive activities. This stage has been known to continue for months.
4 Illumination (right-mode). This is where, normally quite suddenly, a solution may surface: the 'Ah-ha!' experience. This stage, which may be revealed as an image, is frequently associated with relief and pleasure.
5 Verification (left-mode). Although the previous stage may be interpreted as just good luck, the outcome is accepted and the investigator proceeds to test the solution(s) against known information.

This preamble is intended to identify the cognitive, information-processing environment of map use. Effective mental imagery, especially of subject fields comprising very large datasets (such as the Earth's surface), may not be fully accessible to the scientific investigator. In other words the conscious creation and analysis of a distinct 'mind's-eye' image may not always comprise the whole investigative process. Imaging may progress flexibly, holistically and in parallel, but be hidden from the familiar mental working space, and only emerge at the illumination stage. This model, therefore, suggests that creative insights are not the accidental events claimed by some. Shepard (1978) identifies two types of image-inspired creativity. The first is seen in visible 'inventions' and consists of concrete, perhaps metaphorical, externalisations of what were mental images, e.g. the alternating current motor (Nikola Tesla), and special relativity (Albert Einstein). The second is used by both artists and scientists through drawing, painting and so on. It can be seen as attempts to capture ephemeral 'illumination-stage' images that may help solve ongoing problems such as Kekule's sketches of dancing chains of atoms imagined while developing the theory of molecular structure. McKim (1972) offers many other examples, including making physical models. Many researchers use cartographic techniques in their enquiries. Mapping has inspired first insights, provided vehicles for saturation research, and, in some individuals, initiated illuminatory incubation periods! Maps are also highly suited to the verification process, where spatial hypotheses can be tested.

... SCIENTIFIC VISUALIZATION WITH TRADITIONAL MAPS

Because ViSC has been defined to include generalised mappings of data as well as the currently dominant concept of interaction, the role of maps as visualization tools will be examined under both headings.

CARTOGRAPHY AS GENERALISED MAPPINGS

The earliest applications of many basic visualization techniques (sometimes referred to as 'generic' in ViSC literature) may have been cartographic. In fact some books on the subject of ViSC read, in part, almost like veiled cartography texts, with reference to:

(a) Empirical models from data such as the collection of sampled height data from the landscape, the application of interpolation processes and, perhaps, transformation of scale and base geometry via map projection.
(b) Schematics to depict the models such as selecting the idea of a contoured surface to represent the above information.
(c) Rendering the graphic display by, for example, selecting and applying hypsometric layer-tinting to depict what is a representation of the surface of the landscape.

ViSC depictions, like mental images, should be schematic, highlighting key attributes to assist visual processing. Abstraction, or generalisation, is also a fundamental characteristic of cartography, and includes data selection, classification, simplification and symbolisation. Although transformations of map geometry (map projections) boast a long history and extensive literature, cartographic generalisation provides more problems and is less well understood. Complicated by the frequently conflicting criteria of map scale and intended map use, it is one of the most troublesome processes in map-making and continues to resist attempts at computerisation (Buttenfield and McMaster, 1991).

The rendering (or graphic design) stage of ViSC has, perhaps, its earliest provenance in cartography. Although coloured manuscript maps of great beauty can be traced back thousands of years, this discussion will be restricted to the era of printing. Other than text, maps were among the first printed images. This rich historical record, especially of thematic mapping (Robinson, 1982), includes early examples of many modern information-graphic symbols, such as arrow-lines (e.g. Halley's 1686 map of trade winds), isolines (e.g. Halley's 1701 map of magnetic declination), variable shading (Ritter's 1806 physical map of Europe) and colour printing (from mid-nineteenth century). The use of colour predates the latter as engraved maps were being hand-coloured at least three centuries earlier.

By the mid-twentieth century most (carto)graphic symbol techniques were well established. The following shows how generic ViSC terms have cartographic examples:

- Point data: e.g. dot maps.
- Scalar entities of 1D and 2D: e.g. contouring; shading; perspective image display (either smooth-surfaced or with distributions such as geology); multiple-scalar fields such as population pyramids distributed over a map; the use of light and shade to enhance user perception of surfaces (as in hillshading on Swiss topographic maps); bounded-region plots (such as choropleth maps of population density).
- Scalar entities of 3D: e.g. volume visualizations depicted from many viewpoints, with or without surface distributions or enhancements.
- Vector entities of 2D and 3D. e.g. arrow-plots.

Despite this rich array of graphic examples, systematic writing about symbols and map design was slower to emerge. A growing interest in cartography, especially among academics, produced an expanding list of publications from about 1950. Although these were welcome additions there was still a paucity of the recorded thoughts of 'real' cartographers. Shepard (1978) observed that especially creative people might be so right-brained or non-verbal that nothing could be more inimical to their whole approach than to try to articulate it in words. Fortunately, some authors, such as Imhof (1982), Robinson (1952, 1967) and Keates (1982, 1989), proved sufficiently whole-brained to offer, from personal experience, rich and comprehensible explanations of many cartographic principles and procedures. The 1960s also saw a growth in design/perceptual studies of map symbols, mainly by non-cartographers (Wood, 1990), offering a useful resource for designers of map-like graphic images (e.g. Cuff, 1972, 1973, 1974; Doslak and Crawford, 1977). In geographical studies, as in medical visualization, surfaces of interest, for example height or temperature, may need to be segmented using variable parameters. Common requirements such as these have led to an extensive, partially empirical, knowledge-base of the application of graphic variables. With respect to the use of colour, Tufte acknowledges this source in his declaration that 'we have just to figure out what the map people have been doing, and follow that guide' (Brewer, 1992).

Map design, however, is frequently more complex than the mere rendering of a visualization object in ViSC. It embodies much more than the selection of appropriate symbols or techniques. Skill is required to solve the problems of displaying multiple overlays of map content such as contours, woodland, settlement, communications and names. Sadly, these design processes cannot be reduced to a series of simple rules. Some devices, applied with skill, are effective. The visual stratification method (hierarchies) may have emerged as a practical solution for the combination of numerous data layers. This application of the figure-ground phenomenon is now a guiding rule in map design (Dent, 1990), but Castner (1979) has observed that 'about the only real design rule ... is to ensure the clarity and legibility of all map elements'. As the geographical locations of most map symbols are inflexible, the real art of cartography may comprise not only sensitive selection of symbols and application of graphic variables but also the ability to make the thousands of permitted minor adjustments (both locational and graphic) required for a clear, legible and usable image. Defined in this way, Keates (1982) claims that, 'the art of cartography ... is not simply an anachronism surviving from some pre-scientific era; it is an integral part of the cartographic process', and a high-quality map is one of Tufte's standards of excellence in information design (Tufte, 1990). There seems little doubt that traditional maps can satisfy the significant, if subordinate, presentational requirements of ViSC. McKim (1972) describes this process of graphic communication as 'slow and deliberate' where design is critical because the product must satisfy many users.

MAPS AS SOURCES OF INSIGHT

In cartography, the last two decades have been dominated by the growth of cognitive research into mapping and map use (Wood, 1990). Earlier studies of symbol perception were often too mechanistic. Narrow comparisons of cartography with information transmission theory (giving rise to the phrase cartographic communication), while providing scope for intellectual activity, offered little to advance practice of the subject. The links with psychology have been forged and cognitive processes now head the research agenda. While the broadly useful concept of cartographic communication remains, it can still lead to confusion when discussing map use. The distinction has been made between cartographic communication, where an optimal map is designed to communicate a specific message, and cartographic visualization, where the message is unknown and there is no optimal map. Only the latter, it is claimed, can assist users to discover patterns and relationships in geographical data. The existence of many, if any, of the former map type is doubtful, although one example might be a geographer presenting a single proposal in a very simple way. Most maps, therefore, must be suited to some form of cartographic visualization. Analysis of real map use reveals that what the user extracts from a map always differs, at least in part, from the cartographer's input. This is referred to by Robinson and Petchenik (1976) as the 'unplanned increment'. Thus any map, including attempts at a simple message, can be used as a tool in ViSC, offering sources of insight.

Because of the vastness and complexity of the total geographical environment, it is obvious that maps can provide views which are impossible by other means. The tool-like nature of maps is readily accepted by geographers and others, whose uses embrace not only reading but interpretation and analysis. Muehrcke (1981) summarises common analytical techniques as 'cartometry' including measurements on/from maps, map comparison, a holistic awareness of density, arrangement, trends, connectivity, hierarchy and spatial associations. He also observes that some spatial data may have to be transformed into non-operative space. Time–distance, for example, is often important in the study of human geography.

Topographic maps are themselves potential scientific visualization tools. The earth's surface can be regarded as an empirical model from which survey data are merely sampled. This 'visualization object' was originally rendered on printed maps using hachures, but other techniques, including contours, layer tints and hill-shading, appeared later. Geographers use the topographic map of Britain, for example, to investigate patterns of deglaciation through study of landforms. In areas of uncertainty or anomaly, the researcher will seek further more detailed survey data. The addition of other layers, such as roads, vegetation, settlement, geology or soils, not only expands the dataset but increases the potential for insights arising from spatial correlations. Again, in Tufte's words, 'high density designs allow viewers to select, to narrate, to recast and personalise data for their own uses' (Tufte, 1990). Data-thin displays provide little scope for pattern-seeking.

...MAPS AND MAPPING

The previous section assumes the investigator to be using the map as an existing, prescribed, exploratory environment. It is, in a sense, a schematic surrogate, a source of generalised data. But, using McKim's terminology (1972), this 'graphic communication' should be contrasted with 'graphic ideation', which incorporates seeing and imagining. Here, quick freehand drawings may be created as part of the evolving enquiry for personal consumption, and design considerations are not critical. Taken literally, this method, popular with architects and designers, is perhaps less common among map users. Effective scientific visualization through traditional maps may derive something from both approaches. Maps have certainly been used for centuries, not only passively to store the results of surveys, but also actively, to help investigate the complex distributions and spatial relationships, physical and human, of the earth. Real research with maps is never trivial. Data, collected in the field or transferred from other sources, must then be painstakingly plotted on to a base projection. Some simple design rules should apply, even if the product is not destined for publication, because, without clarity and legibility, potential insights may go undetected. A geographer may display a given set of data using a variety of techniques (e.g. isoline, choropleth), at various resolutions and with a range of classification steps and colour sequences, to stimulate the pattern (insight) seeking process. This is far removed from McKim's 'quick freehand sketch' but may function in a similar manner.

The history of thematic mapping, that is the mapping of distributions other than topography, is bound up with scientific investigation and research. In the early nineteenth century medical practitioners were preparing innovative maps, normally in manuscript form, as research tools. Many hoped to identify environmental factors which might correlate with the occurrences and spread of disease. Similarly, Brandes, when mapping the patterns of wind direction in a 1783 storm, revealed the now familiar relationship between wind and atmospheric pressure, undetectable in the tabulated data. Yet again, Guettard, while doing exploratory mapping of plants, observed spatial relationships with minerals and endorsed the important contributions made by such mapping to general theoretical studies (Robinson, 1982).

... CONCLUSION

The evidence presented appears to substantiate the claim that traditional maps have been strikingly effective tools for scientific visualization. Casual users, observing them as pictures, may be as unaware of their insight-potential as they would be of the insight to be derived from a computer-generated (ViSC) image. The value of interaction with map data was well appreciated in the past. Scientists, using cartography, knew that having a limited set of available representations could limit the ways that they thought about their models and thereby limit potential insights (Palmer, 1992). Geologists, for instance, seeking further information about their 3D domain, produced cross-sections or depicted underground surfaces (e.g. salt domes) to help view structural relationships. Similarly, employing one of the most useful principles of modern medical visualization, map-makers created block diagrams to analyse surface data from other viewpoints. The general problem in the past was execution time. These manual methods often took too long to be useful in urgent analyses, although this time factor may have been less critical with the slower pace of earlier scientific research.

The interactive process can be interpreted in another way. The map provides an information source which can offer up the makings of many unexpected patterns from which insights may emerge. As even the most complex maps contain limited input data, when compared with the real world and do not have simply defined messages, observed patterns must be the result of an interactive cognitive dialogue between the map and the viewer's mind. The perfect visual working environment for this process is data-rich and well designed. In Tufte's words (1990) 'the more relevant information within eyespan, the better'.

While there is little doubt that the animation and high-speed interaction available from new ViSC systems can vastly improve the rate of investigative research in certain fields, what we are seeing is just the latest stage in a continuum of technique development. ViSC (including 2D and 3D GIS) may offer a quantum leap in tool sophistication but, even today, more leisurely study of high information, traditional-style maps can provide scientific insights of great value. More importantly, however, the emergence of integrated databases, computer assistance, multimedia and ViSC is leading to an expanded and more

15

powerful 'new' cartography. The cartographers of the past creatively explored the bounds of their discipline within the restrictions of contemporary data collection, manipulation and production methods. They used 3D views of data, integrated plans and cross-sections, animation, etc., but often with extreme difficulty. Perhaps their successors, the modern practising 'traditional' cartographers, have not been pioneers in exploring the current computer-related possibilities of their discipline. This does not mean that cartography has not advanced. As often happens, new technologies and theories can create diversions. Cartography as a concept is currently overshadowed by GIS, a facility which is certainly transforming the realms of geographically related data storage, integration, manipulation and access. There is no doubt that GIS (and similar systems in other scientific fields) can provide answers to certain questions very effectively without the use of maps or map-like graphics. The other viewpoint is, however, that maps, including visualizations of all kinds generated from such systems can be so much more effective than those in the pre-GIS past. Scientists of many kinds have always appreciated the value of maps. Because of their rapidly expanding needs to handle oceans of new datasets and their close involvement with pioneering computer-related technologies, it is they who have, inadvertently, been the pioneers of 'new' cartography. To quote Michael Zeitlin (1992) of Texaco:

> it was impossible to look at all the (geological) data at once before the advent of high speed parallel processing, 3D graphics and ... visualization. Today we can grasp enormous amounts of information ... In other words we see – and use – more of our data and process it with visual pattern recognition as the basis of our interpretation.

The term Scientific Visualization may have been coined to describe this new era of graphic discovery but it still remains a form of mapping. Indeed any such new advances in the graphic display of information will, inevitably, remain subsets of cartography. In the words of Hall (1992: 'The widespread availability of computers with specialised graphics software puts the equivalent of a cartographer inside every computer, and therefore inside every laboratory group.'

... REFERENCES

Brewer C A (1992) Review of Envisioning Information by Tufte E R. *Photogrammetric Engineering and Remote Sensing* LVIII(5): 544–545

Brodlie K W, Carpenter L A, Earnshaw R A, Gallop J R, Hubbold R, Mumford A M, Osland C D, Quarendon P (eds) (1992) *Scientific Visualization: Techniques and Applications*, Springer-Verlag, Berlin

Buttenfield B, McMaster R B (eds) (1991) *Map Generalization: Making Rules for Knowledge Representation*, Longman, Harlow

Castner H W (1979) Viewing time and experience as factors in map design research. *Canadian Cartographer* 16(2): 145–158

Cuff D J (1972) Value versus croma in color schemes on quantitative maps. *Canadian Cartographer* 9(2): 134–140

Cuff D J (1973) Shading on choropleth maps: some suspicions confirmed. *Proceedings, Association of American Geographers*, Annual Meeting 24 (April): 50–54.

Cuff D J (1974) Impending conflict in color guidelines for maps of statistical surfaces. *Canadian Cartographer* 11(1): 54–58

Dent B D (1990) *Cartography, Thematic Map Design*, WCB Publishers, Dubuque

Doslak W Jr, Crawford P V (1977) Color influence on the perception of spatial structure. *Canadian Cartographer* 14(2): 120–129

Edwards B (1987) *Drawing on the Artist Within*, Collins, London

Flynn G W (1990) Chemical cartography: finding the keys to the kinetic labyrinth. *Science* 246: 1009–1015

Haber R B, McNabb D A (1990) Visualization idioms: a conceptual model for scientific visualization systems. In: Neilson G M, Shriver B, Rosenblum L J (eds) *Visualization in Scientific Computing*, IEEE Computer Society Press, pp. 74–93

Hall S S (1992) *Mapping the Next Millenium: The Discovery of New Geographies*, Random House, New York

Hamming R W (1962) *Numerical Methods for Scientists and Engineers*, McGraw-Hill, New York

Imhof E (1982) *Cartographic Relief Representation*, Walter de Gruyter, Berlin

John E R, Prichep L S, Fridman J, Easton P (1988) Neurometrics: computer-assisted differential diagnosis of brain dysfunctions. *Science* 239: 162–169

Keates J S (1982) *Understanding Maps*, Longman, Harlow

Keates J S (1989) *Cartographic Design and Production* 2nd edn, Longman, Harlow

McKim R H (1972) *Experiences in Visual Thinking*, Brooks/Cole, Monterey

Muehrcke P (1981) Maps in Geography. *Cartographica* 18(2): 1–41

Palmer T C (1992) A language for molecular visualization. *IEEE Computer Graphics and Applications* 12(3): 23–32

Robinson A H (1952) *The Look of Maps*, University of Wisconsin Press, Madison

Robinson A H (1967) Psychological aspects of colour in cartography. *International Yearbook of Cartography* VIII: 50–61

Robinson A H (1982) *Early Thematic Mapping in the History of Cartography*, University of Chicago Press, London

Robinson A H, Petchenik B B (1976) *The Nature of Maps*, University of Chicago Press, London

Shepard R N (1978) Externalisation of mental images and the act of creation. In: Randhawa B S, Coffman W E (eds) *Visual Learning, Thinking and Communication*, Academic Press, London

Tufte E R (1990) *Envisioning Information*, Graphics Press, Cheshire Connecticut

Wood M (1990) Map perception studies. In: Perkins C R, Parry R B (eds) *Information Sources in Cartography*, Bowker-Saur, London, pp. 441–452

Zeitlin M (1992) Visualization brings a new dimension to oil exploration and production. *Geobyte* 7(3): 36–39

3 Visualisation in GIS, cartography and ViSC

M. Visvalingam

... Introduction

This chapter explores differing perspectives on visualisation. First different meanings of the word are distinguished. Then, the role of visual analysis within geographic enquiry and some current capabilities of ViSC systems are examined. Next, different perceptions of the fields of GIS and cartography are considered. Developments in technology are providing new opportunities and have led to the establishment of the discipline of ViSC. The rapid cross-fertilisation of ideas across these fields is focusing attention on inter-disciplinary issues. From a technological perspective, digital cartography, like computer-aided design, has been viewed as an application of computer graphics. Similarly, cartography and GIS could become merely applications of ViSC. The chapter, therefore, goes on to examine some reasons why current ViSC systems may not be adequate for them. The chapter concludes with the suggestion that all three disciplines should look beyond the generation and manipulation of pictures and investigate concepts, techniques, tools and systems for exploring and understanding data through a combination of images, tables and other representations.

... Visualisation and visualization

The word visualisation has acquired different interpretations. It is defined (Oxford English Dictionary) as 'the power or process of forming a mental picture or vision of something not actually present to the sight' or, as a noun, 'a picture thus formed'. The term is also used to refer to the process of making a visible image of something otherwise invisible, as in X-ray photography. Within the field of ViSC, the term visualization (spelled with a 'z') is a synonym for ViSC and has acquired two further connotations. First, the term visualization represents the use of computer technology for exploring data in visual form and for experiencing virtual worlds using all the human sensory channels. Second, it focuses attention on the use of computer graphics for acquiring a deeper understanding of data. The capacity to generate and interact with realistic images has been exploited within advertising, flight simulation, medical imaging, environmental impact analysis and many other applications. ViSC seeks to exploit this technology to provide tools, techniques and systems specifically for invoking insight in science and engineering. To avoid confusion, I will assign specific meanings to the following terms in this chapter:

Visualisation is primarily a mental process which serves a variety of purposes, including visual analysis. Visual analysis refers to the use of visualisation as a distinct method of inquiry for provoking insight and for concept refinement.

ViSC is the discipline concerned with developing the tools, techniques and systems for computer-assisted visualisation. 'It studies those mechanisms in humans and computers which allow them in concert to perceive, use and communicate visual information.' (McCormick et al., 1987, p. 3) In mental visualisation it is difficult to distinguish between the process and product of visualisation since the latter tends continually to mirror the flow of imagination. When visualisation is externalised we can make a distinction between the process and the product.

Visualization refers to the process. It is 'a series of transformations that convert raw simulation data into a displayable image. The goal of the transformations is to convert the information into a format amenable to understanding by the human perceptual system.' (Haber and McNabb, 1990, p. 75)

Visual representations refer to the products, namely the pictorial depictions of mental imagery and/or of data in any medium.

Visual display refers to transient and easily modifiable visual representation on electronic media, such as CRT screens and LCD displays.

Note that in the rest of this book the term has been spelled 'visualization' without making the distinctions I develop in this chapter.

... VISUALISATION IN GEOGRAPHIC ENQUIRY

The aims of ViSC are not new to geographers. Much geographic enquiry is rooted in empiricism. It seeks to discover and provide explanations for the patterns and relationships which exist and the expectations which are violated in our physical and human environment. Geographers have traditionally used visual analysis because the display of data within a spatial framework enables us to recognise patterns almost instantly.

Cartographic maps, based on the geographical framework, have been widely used in geographic analysis. Maps reduce our worlds of enquiry into assimilable proportions and cast them into shapes from which we can derive information. The common frameworks for visual cross-referencing and for linking of data on different map coverages enable us to use our personal stores of loosely structured geographical knowledge during interpretative stages.

GIS are tools for quantification within the paradigm of 'geography as geometry'. Muehrcke (1981, p. 3) described the latter as a 'philosophy based on map-like spatial measures of distance, shape, direction, area and so forth'. The proponents of GIS tended to focus on the data management and analytical functions of GIS. This has encouraged a myopic view of visualisation as just one method for communicating results. The US National Science Foundation report on 'Visualization in Scientific Computing' (McCormick et al., 1987) was largely responsible for reiterating the value of visualisation to a wider scientific community and for re-focusing attention on the need to develop and exploit ViSC.

... COMPUTER-ASSISTED VISUALIZATION

Computer graphics enable us rapidly to display the vast volumes of data processed using GIS which we would not consider doing manually. For example, our capacity for generating mental images of three-dimensional objects, let alone their relationships, is limited. Computers can produce these things for us. Geographers have exploited mapping packages since the 1970s for modelling and viewing physical, demographic and socioeconomic variables. Many GIS now facilitate the generation of 3D views of landscape, overlaid by land use and other cultural information, and of visual representations of multidimensional data. ViSC systems have also extended our capacity for volume visualization. Systems developers and researchers are competing with each other to find more and more ways of displaying multidimensional data (see Brodlie et al., 1992).

Both GIS and ViSC systems have taken advantage of the achievements of computer graphics. However, the speed of response of modern computers and developments in graphical user interfaces (GUIs) have been exploited within ViSC in a variety of imaginative ways and to a larger extent than within GIS at present. Within ViSC, a GUI promotes interactive visual programming using dataflow diagrams. A network editor enables the user to build applications by selecting data and techniques, both represented in iconic form, and wiring (connecting) their input and output ports. The user can merge and fork the streams of data flowing through the network and thereby project multiple perspectives on data. The GUI allows the user to fine tune the parameters driving various procedures by manipulating physical interaction devices and/or virtual devices, called widgets (window gadgets). In a visual programming environment, the user can identify appropriate network configurations and parameters to simplify, model and image the data. Visual programming, network editors and widgets promote experimentation but the underlying conceptual model of the process of visualization of data is not substantially different from the model underpinning cartographic and GIS software.

Animation is useful for exploring data and processes (see Chapters 13 and 14). For example, some properties of the Douglas-Peucker algorithm for line simplification, which had escaped notice for some twenty years, were noticeable when its processing of lines was animated (Visvalingam and Whyatt, 1991). The speed of modern computers facilitates interactive manipulation of the objects being visualised in real time. It is possible to zoom, pan, rotate, slice and penetrate objects and scenes using interactional devices such as data gloves and spaceballs (Brodlie et al., 1992). This class of operations facilitates visual browsing of data.

There is also scope for interactive probing of data. Data probes are virtual input devices for pointing at data objects to read off data values. Since visualisation is ultimately a human process, human factors should dominate the design of visual representations (see Chapter 21) and the user interfaces to computing systems (see Chapter 22). Graphical interaction and visual programming should in theory lead to more usable systems. But as Rosenblum and Nielson (1991, p. 16) pointed out, current visualization systems still require much effort to use. Many end-users, including scientists, may need to rely on the services of 'vizineers' (cognitive engineers with expertise in visual processing). As they stated: 'Perhaps the most important need today is for truly usable but still powerful and extensible tool sets.' (p. 17)

... ORIENTATIONS OF GIS AND CARTOGRAPHY TOWARDS VISUALISATION

Developments in information technology are offering GIS, cartography and ViSC new opportunities for growth and overlap. There is a tendency to resolve the resulting confusion by comparing GIS and ViSC in terms of the capabilities and features of their state-of-the-art proprietary systems but, given the rapid rate of increase of ideas and methods, the current features of such systems cannot be assumed to define the total scope of their disciplines. There is, however, a need to understand the specific orientations of these three disciplines so that we can identify the significance of their different perspectives and concerns.

As stated by Coppock and Anderson (1987), GIS is as old as cartography and predates computers. As it evolved to take advantage of pertinent developments in computer-based technology, the emphasis has shifted from the data management to the analytical functions. The Chorley Report (DoE, 1987, p. 132) defined a GIS as 'a system for capturing, storing, checking, integrating, manipulating, analysing and displaying data which are spatially referenced to the Earth. This is normally considered to involve a spatially referenced computer database and appropriate applications software.'

Despite its wide and disparate concerns, cartography has a well-defined and unique focus, namely maps (Visvalingam, 1990; see also Chapter 2). Despite this well-defined focus, the perception of the role of digital cartography has also altered. Initially, it was seen as interacting with GIS in one of three ways 'by providing a framework, through national map coverages, for relating other categories of data; as a source of spatially referenced data in its own right; and as one of the methods for presenting the results of analyses of such data' (Coppock and Anderson, 1987, p. 4). The data storage and dissemination functions of the traditional map are being subsumed by spatial databases, which are fast becoming the ultimate digital reference maps. Routine applications of GIS rely on digital, rather than visual, models for cross-referencing and analysing data (Visvalingam, 1990).

Within GIS, the visual map is seen mainly as a device for communicating the results of 'What if?' analysis. Visual maps can also be used in a routine way when methods of production and interpretation conform to reliable and well-established practice. Business graphics in management reports and the run-of-the-mill atlas maps usually fall into this category. However, with the advent of high performance personal workstations with enhanced graphics facilities running UNIX, it has been obvious for some time that visual maps could themselves become two-way virtual devices for probing and exploring the digital map (Visvalingam and Kirby, 1984; Visvalingam, 1989). Muehrcke (1981, p. 37) warned that followers of the 'geography as geometry' ideology were failing to take advantage of 'the powerful insight-generating and communication potential of cartographic abstraction' and pointed out that 'words and numbers and images are complementary, not substitute, vehicles of thought'.

Muehrcke (1981) also noted that the emergence of professional cartography has led to a separation of the processes of map-making and map use. This is perfectly reasonable when maps are used as inventories for information communication but is inappropriate when maps are used as vehicles of thought since the purpose for which a map is to be used is the primary determinant of its design. He anticipated that: 'With a little assistance from geographic cartographers, and with the examples provided by a diversity of other disciplines which are discovering the importance of visualization in science and art, geographers themselves may soon be born-again map users.' (p. 51) The evangelistic role of ViSC is welcome but my main concern is that it, like professional cartography, appears to be placing the emphasis on construction, rather than flexible use, of visual displays.

...GIS REQUIREMENTS AND ViSC SYSTEMS

Both GIS and digital cartography have taken advantage of developments in computer graphics but not necessarily as bundled within ViSC systems. There are several reasons for this.

First, ViSC systems are structured to pre-process (filter/enrich/manipulate), map and render data (see Chapter 6). As M. Wood (see Chapter 2) points out, this model of visualization is not dissimilar to the original approach adopted within digital cartography. The parameter-driven approach, as exemplified by command driven mapping packages, is still favoured by some users. When the need for imaginative design is limited, automation is facilitated. CAD packages and paintbox systems, based on object-oriented programming, have proved to be much more convenient for artistic visualization. The scope for direct manipulation within these systems has enabled cartographers to practise computer-assisted, near-manual cartography on the versatile electronic screen, where they can easily modify software-generated displays. ViSC systems are not designed to facilitate direct modification of images.

Second, cartography and GIS place great importance on the origins and nature of input data. It is generally known that available data may be inadequate with respect to accuracy, currency, completeness and structure (see Chapter 16). Thus, the need to validate, edit, update and restructure data have led to the inclusion of intelligent graphical editors within cartographic and GIS software. Maps are not just devices for presenting and communicating data, they are also used for massaging and managing both spatial and associated aspatial data. At present, ViSC does not seem to address these mundane requirements. No doubt, some of the requirements for map design and digital mapping can be met by pre- and post-processors. Within ViSC systems, there is ample scope for setting control data through widgets but it is not easy to modify the data being visualised by direct manipulation of the visual representation.

Third, GIS use complex data structures for linking the spatial and the aspatial descriptions and attributes of geographic entities. Relational and, increasingly, object-oriented models are used for accessing the aspatial data. The spatial descriptions, on the other hand, are manipulated by a separate subsystem. The name of the Arc/Info GIS reflects this dual processing of data. Spatial data models also need to accommodate the added dimensions of time and scale. The characteristics of spatial data are very different from those of image and volume data and spatial data modelling remains a prominent on-going topic of research (Frank, 1991). ViSC systems cannot fully cope with GIS data at present.

Fourth, different representations of the same data or of a process enable us to focus on different aspects of the problem. This is well appreciated within systems for algorithm visualization. In geographic research, understanding has been furthered by comparing distributions of both individual and aggregate elements in a set of maps in conjunction with use of tables of raw and processed statistics. Semi-manual cross-referencing and exploring of multivariate data was laborious even when any spatial statistics related to a relatively small set of data collection units. Computer technology has facilitated the collection, statistical analysis and display of data for a very large number of units of much higher resolution. Yet the older GIS and current ViSC systems have not encouraged visual cross-referencing and exploration. This may be partly because of a lack of appropriate models for interaction and partly because of inappropriate metaphors for packaging software. The idea of using data displays as virtual devices for retrieving and/or highlighting spatial and other related data in complementary views is, however, already present to some extent within some recent GIS.

Fifth, the main thrust of computer graphics has been towards making the display screen into a window through which one beholds a virtual world. This inspired a programme of research towards the simulation of realism through concepts and procedures for geometric modelling of 3D objects and scenes. These include hidden-surface removal, surface texturing, lighting, movement, volume visualization and virtual worlds. No doubt there are GIS applications, such as environmental impact analysis, geologic exploration and military uses, which already benefit from photorealism, volume visualization and virtual reality (see Chapter 8). However, the so-called 'natural scene' paradigm (Robertson, 1991) is not appropriate for many applications. It is not just the limits of technology nor of human capacity which discourage us from representing reality in all its complexity. We simplify in order to focus attention on the important, as opposed to the irrelevant. Simplification often involves a reduction in the number of dimensions, the scale of measurement or of categories, and a generalisation of forms. While there have been considerable advances in the achievement of realism, there has not been corresponding progress in automation of map design and generalisation. Our inability to devise algorithms which can mimic the cartographer's art implies that despite nearly thirty years of research we do not yet understand the cartographer's internal visualisation.

Much effort is being expended on formalising already externalised knowledge and rules, inherited from traditional cartography, within knowledge based systems (Buttenfield and McMaster, 1991). Such declarative knowledge is probably insufficient. What we need to know is when cartographers adhere to these rules and how and why they either bend or ignore rules and instead exercise artistic licence and judgement. It is ironic that while scale-free mapping remains a long-cherished and still unrealisable goal, skilled cartographers have been redeployed as map digitisers and operators of command-driven mapping systems. Fortunately, the wide availability of paintbox type systems is now providing opportunities for cartographers to continue the development of their art and skills within digital environments. One of the priority areas for research in virtual environments is the creation of synoptic overview maps of 3D data. This requires that the data model is structured such that the important features in the model are separated from the detailed description of shapes.

Finally, experts in ViSC believe that the dataflow model of the process of visualization, as presented in Brodlie et al. (1992), is sufficiently flexible to accommodate and articulate visual analysis in geography. It is a well-established programming paradigm which provides a convenient implementation model for picture generation, but it does not seem to be an appropriate user model. ViSC systems are already too complex for most end-users. Upstream flows of data and additional modules for transforming them are likely to create spaghetti diagrams. ViSC is a relatively new field and we are likely to see many new ideas. It is therefore quite likely that GIS vendors will develop new metaphors for constructing visualization systems for interactive graphical investigation of spatial data.

... CONCLUSION

GIS, cartography and ViSC are complementary disciplines. GIS and ViSC have become constrained to exploiting computer technology. The scope of cartography, on the other hand, is not limited by computability since its primary concern is the exploitation of human vision, preferably, but not necessarily, with computer assistance. Visual representations are used to manage, access, analyse and view complex digital models of reality. They range from annotated images of reality, as in photomaps, to cartographic abstractions, such as line maps and cartograms. They capture and communicate reality, and provoke specific interpretations and insights. Their construction and use may be prescriptive and formula-driven, as in many well established and routine applications; or they may be hypothesis-driven and form part of a process of inquiry and discovery.

McCormick et al. (1987) stated that ViSC studies those mechanisms in humans and computers which allow them, in concert, to perceive, use and communicate visual information. However, its past emphasis has been on the development of techniques, tools and systems for invoking insight through manipulable visual representations. Already, ViSC has extended our capacity for visualising phenomena in three and higher dimensions. While the advocates of quantitative geography and GIS have tended to underestimate the value of visualisation, the proponents of the new discipline of ViSC are perhaps inclined to exaggerate it. All methods of inquiry, whether textual, mathematical or visual, have their own advantages and limitations. They are all open to over-zealous promotion and to inadvertent or deliberate misuse. Visual analysis is no exception.

... References

Brodlie K W, Carpenter L A, Earnshaw R A, Gallop J R, Hubbold R J, Mumford A M, Osland C D, Quarenden P (eds) (1992) *Scientific Visualization: Techniques and Applications*, Springer-Verlag, Berlin

Buttenfield B P and McMaster R B (eds) (1991) *Map Generalization: Making Rules for Knowledge Representation*, Longman, Harlow

Coppock J T, Anderson E K (1987) Editorial review. *International Journal of Geographical Information Systems* 1(1): 3–11

Department of the Environment (DoE) (1987) *Handling Geographic Information*, HMSO, London

Frank A U (1991) Properties of geographic data: Requirements for spatial access methods. In: Gunther O, Schek H-J (eds) *Advances in Spatial Databases: Proceedings of the 2nd Spatial Data Handling Symposium*, Zurich, pp. 225–234.

Haber R B, McNabb D A (1990) Visualization idioms: a conceptual model for scientific visualization systems. In: Nielson G M, Shriver B, Rosenblum L J (eds) *Visualization in Scientific Computing*, pp. 74–93.

McCormick B, DeFanti T A, Brown M D (eds) (1987) Visualization in scientific computing. Special issue ACM SIGGRAPH *Computer Graphics*, 21(6)

Muehrcke P (1981) Maps in geography. *Cartographica* 18(2): 1–41

Robertson P K (1991) A methodology for choosing data representations. *IEEE Computer Graphics and Applications* 11(3): 56–67

Rosenblum L J, Nielson G M (1991) Guest editor's introduction: visualization comes of age. *IEEE Computer Graphics and Applications* 11(3): 15–17

Visvalingam M (1989) Cartography, GIS and maps in perspective. *Cartographic Journal* 26(1): 26–32

Visvalingam M (1990) Trends and concerns in digital cartography. *Computer-Aided Design* 22(3): 115–130

Visvalingam M, Kirby G H (1984) *The Impact of Advances in IT on the Cartographic Interface in Social Planning*, Department of Geography Miscellaneous Series 27, University of Hull, Hull UK

Visvalingam M, Whyatt J D (1991) Cartographic algorithms: problems of implementation and evaluation and the impact of digitising errors. *Computer Graphics Forum* 10(3): 225–235

4 Cogent GIS visualizations

A. Turk

Geographic information systems are becoming more flexible, powerful and sophisticated and more closely integrated with modelling software. Hence, the communications between users and GIS are increasingly interactive and complex, especially when GIS are configured as spatial decision support systems (SDSS). Developments in GIS visualization techniques have been driven by these factors, as well as by improvements in analysis and display functionality.

Visualization is a very active field of research and development but the term means different things to different people. Some authors incorporate within it all graphics aspects of communication between computers and people. Traditionally in the field of GIS, visualization has tended to have a much more restricted usage, usually associated with conventional maps or three-dimensional perspective views of terrain or statistical surfaces. In more recent times the term has also been used to cover graphics representing projected environmental or land-use change which may effect the visual amenity of a locality (Smart and Mason, 1990; Bishop and Hull, 1991). The meaning of visualization in the GIS context is now even broader, including, for example, representations of data quality (Buttenfield and Ganter, 1990).

GIS visualizations are becoming increasingly sophisticated. They may display relationships between conceptual as well as physical entities, be two-dimensioned (van Elzakker, 1991) or three-dimensional (Kraak, 1991), and may involve image draping and stereoscopic viewing (Sarjakoski, 1990; Moellering, 1991). As well as this trend towards more realistic visualization, there is increasing use of abstract graphics to represent the distribution of non-visual phenomena and to describe processes (Monmonier, 1990; MacDougall, 1991). Applications of this type often utilise sequences of graphics/images to provide an animated message, especially to communicate changes in phenomena over time, for example, in global change research (Stephenson, 1990; Goodchild, 1991a; MacEachren, Chapter 13 of this volume). The design and use of animated cartographic visualizations to depict spatio-temporal distributions of geographic phenomena is discussed by Dibiase et al. (1991) and by Openshaw, Waugh and Cross (Chapter 14, this volume). They review developments in dynamic mapping techniques over the past three decades, including 2D and 3D approaches, and report a map animation project in the epidemiological mapping of AIDS statistics. The lack of design guidelines for such visualizations is noted and they discuss several design constraints flowing from the duration of viewing of each graphic and the hardware used.

...APPROACHES TO THE DESIGN/SELECTION OF GIS VISUALIZATIONS

The choice of visualization content and form may be considered in terms of establishing an appropriate set of relationships between these parameters, the range of possible communication objectives, and relevant contextual variables. The definition of the user's decision-making objectives, and the design and/or selection of GIS visualizations to accomplish them, may be executed at increasing levels of formality:

1 Informal: visualization design/selection by 'experts' on an intuitive and/or heuristic basis (in terms of their particular experience and, possibly, naive implicit theories).
2 Theory based: visualization design/selection by use of an integrated set of explicit theories from relevant disciplines, rendered coherent and tractable through an integrated theoretical model, but not operationalised as principles or procedures.
3 Principles/procedures based: visualization design/selection through the application of a set of principles and rules (governing likely circumstances), formal models and design procedures.
4 Automated: visualization design/selection using knowledge-based system software which implements formal (computational) analysis and design models.

The validity of the implicit assumption that these approaches will be successively more efficient and effective depends upon how well the relevant causal variables are modelled and the suitability of the procedures adopted. It is therefore important that the validity and effectiveness of more formal approaches be evaluated during their development.

Robertson (1991) proposes a formal methodology for choosing data representations. He highlights the range of potential visualization products which may be used to represent any dataset and proposes a 'natural scene' paradigm as the basis for a visualization design/selection procedure. This approach seeks to communicate specific information about a dataset by using realistic visualizations which incorporate recognisable properties of objects and scenes appropriate to the user's existing mental models. The methodology is based on:

> establishing the nature of the data to be displayed and the interpretation aims of the analyst; establishing the capability of various visual representations, as components of natural scenes, to convey information about the attributes of data variables; and choosing an appropriate representation, or set of representations, for the data by matching the representation capability to the interpretation aims. (Robertson, 1991, p. 60)

Robertson suggests that visualization scene properties can convey information about data variables both implicitly and explicitly and that this cognitive transaction may be matched to the user's interpretation requirements via the proposed methodology. Alternative paradigms may be used to suit different data types and communication objectives. Although the reported procedure is a considerable improvement over an ad hoc approach, it needs further development in order to deal effectively with a broad range of abstract and/or dynamic representations and with user interaction in the visualization process.

··· OPTIMISING VISUALIZATIONS THROUGH COGNITIVE TASK ANALYSIS

If GIS visualizations are to be optimised it is necessary rationally to address the relationship between communication objectives and the nature of the display within a user-centred, cognitive ergonomics framework (Turk, 1990, 1992). This will be facilitated by the use of cognitive task analysis procedures. The means–ends structure of any GIS-based decision process defines the cognitive task requirements and the sets of potential mental strategies which may be used. A cognitive ergonomics analysis enables the identification of representation and interpretation requirements. The interaction of these requirements with the viewers' roles and characteristics may be analysed to infer the visualization design parameters. Visualization design/selection may be undertaken as a formal procedure which implements a user-GIS interaction model through a cognitive task analysis procedure. However, a vast array of interaction models and task analysis procedures exist (for example Harrison and Thimbleby, 1990; Olson and Olson, 1990; Ziegler and Bullinger, 1991). A Cognitive Ergonomics in GIS Reference Model can be used as a means of choosing between alternative approaches and tailoring procedures to suit specific aspects of GIS design and evaluation, including the design/selection of visualizations (Turk, 1992).

··· TAXONOMIES FOR VISUALIZATION DESIGN/SELECTION PROCEDURES

A formalism (model) linking visualization objectives and products needs to utilise multidimensional, generic task and form taxonomies. Hence, a necessary aspect of the development of effective visualization design/selection procedures is the development of appropriate taxonomy dimensions and categories. Whether any particular taxonomy dimension is useful or not will depend upon how it reflects the causal factors which dictate the degree of success of a visualization sequence. This question partly turns on the nature of the design optimisation procedure being adopted. Hence, any development of visualization design procedures needs to be embedded within a broader approach to system design methodology. Visualization tasks must form a subset of an overall GIS task taxonomy used to design and evaluate user–GIS interaction.

The purpose of visualizations may be formalised in terms of dimensions such as those suggested by Ganter (1988). These classify visualization graphics by the following broad categories of use:

- Exploratory graphics. Graphics which portray the information generated from numerical simulations or other modelling, especially where there is a need to simplify the presentation or render it less ambiguous or more convincing. 'These graphics usually mimic the appearance of the object or process being studied, and are often dynamic, showing behaviour over time.' (Ganter, 1988, p. 234)

- Design graphics. These graphics are 'an externalization of non-verbal creative thought which permit preliminary testing and comparison of solutions to technological problems' (Ganter, 1988, pp. 234–5). The most common example is graphics in computer-aided design.
- Reference graphics. Graphics of this type 'such as maps, diagrams, and curves are archives of displayed data which can be extracted and put to new uses' (Ganter, 1988, p. 235). They are frequently prepared for a variety of purposes, possibly some considerable time prior to their use, and their accuracy and completeness may be subject to constraints beyond the control, or even the knowledge, of the user.
- Presentation graphics. These are usually simplified graphics designed to communicate specific concepts in a particular context. Their general form may be similar to that of reference graphics.

A visualization designer must determine what phenomena need to be displayed, and the form of the representation, so that the defined communication objectives (cognitive tasks) will be achieved. One taxonomy dimension which may facilitate this process is illustrated by the following list. The different categories are illustrated by examples relating to an environmental pollution application, such as that discussed in Moore (1991):

- Phenomena visualization. Depiction of natural or man-made phenomena, recorded in terms of either point, local or global variables. An example is air pollution readings for specific locations.
- Meta-phenomena visualization. Display of the content/coverage, quality, accuracy, etc., of a particular phenomenon. An example is the accuracy of air pollution values.
- Phenomena change visualization. Depiction of phenomena change over some specific time period, or the rate of change of a phenomenon or one of its attributes. An example is the change in mean annual levels of carbon dioxide in the air for a specific location and altitude between 1980 and 1990.
- Visualizations of relationships between phenomena. Display of specific, spatially based, relationships between phenomena of interest. An example is the pattern of correlations between levels of particular pollution indicators and incidence of asthma in children for a set of locations in a study area.
- Causal visualization. Depiction of cause–effect relationships, known or inferred, involving the phenomena, for example the relationship between winds and air pollution dispersal.
- Meta-causal visualization. Displays of the reliability, validity, etc., of inferred causal relationships, for example air pollution dispersal probability for given wind directions.
- Information system (GIS) structure visualization. Depiction of the information system analysis/display functionality, for example the software modules needed to compare two different datasets.
- Analysis process visualization. Graphic depiction of the processes of analysis used to generate a particular visualization, for example the surface interpolation algorithm used.

- Motivational visualization. Graphic displays designed to catch and hold the viewer's attention, for example a photograph showing severe air pollution over a city.

The visualization intent must be implemented through a specific set of graphics. This requires consideration of another taxonomy dimension covering the form of presentation. An example of the sort of classification of form which may be used is the following list, again with examples from an environmental pollution application:

- Direct display. Uses a 'realistic' visual display to depict a phenomenon which is intrinsically visual, or at least a key aspect under study is visual. An example is the display of simulated smoke pollution levels over a city.
- Indirect display. Graphics/images are used to depict a non-visual phenomenon, where the viewer is consciously or unconsciously aware that the visual display is acting as a surrogate for something real but invisible. An example is a graphic depicting the level of invisible air pollution gases.
- Abstract graphics. In these, the information is rendered in abstract terms, for example a diagram depicting processes causing increased air pollution.
- Metaphorical displays. In these the graphic display is in terms of some (explicit or implicit) metaphor, for example a 'skull and crossbones' to represent poisonous gas.
- Aesthetic graphics. In which the visualizations are designed to produce some emotional response in the viewer, for example a photograph of an asthmatic child.

It is important to note that the visualization design taxonomy dimensions which are appropriate will depend to some extent on the nature of the information to be depicted and the task analysis procedures adopted. In practice, generic visualization task and form taxonomy dimensions may need to be supplemented by dimensions which support the design process in terms of the theory of interaction on which it is based. For instance, a mode of engagement dimension may be appropriate for a task analysis procedure which is based on the 'levels of cognitive control' theoretic model (Rasmussen et al., 1990). For such a dimension, the viewer's mental engagement with the visualization may be considered to be at one of the following levels:

1 Theory/knowledge based. Decision-making by the application of theories and mental models relevant to the visualization sequence.
2 Principles/rules based. Decision-making through the use of sets of principles and rules, triggered by appropriate codes or visual cues.
3 Automated/skill based. Decision-making through automated (skilled) responses to familiar tasks represented in the visualization.

... DESIGNING SEQUENCES OF GRAPHICS

For many applications it is necessary to design visualization sequences as well as individual graphics. Monmonier has coined the term 'graphic script' to refer to a temporally sequenced, multi-window, graphic presentation. 'An individual window might contain a map, text with a definition or numerical statement, or an aspatial statistical graphic such as a histogram or scatterplot. The script controls the information content and symbolization of the display.' (Monmonier, 1989b, p. 381) The script also specifies the size and scale of the graphic elements and their position and duration on the screen. This idea provides a practical procedure for controlling and specifying parameters involved in the design, execution and evaluation of visualization sequences. The script may be a combination of reusable short sequences together with other material particular to the specific communication requirements.

Monmonier (1989) further suggests that 'graphic phrases' may aid script development by acting as standardised sequences into which particular parameter values may be inserted. In this way a certain amount of automation of the process may be possible, provided some mapping is achieved between standard graphic sequences and visualization objectives, determined by cognitive task analysis and expressed in terms of an appropriate taxonomy. The design of graphic scripts should also take into account such psychological parameters as 'visual momentum' and the cognitive characteristics of individual users (Hearnshaw, 1991).

DiBiase et al. (1991b) suggest that viewing time may be utilised as a communication variable in dynamic visualizations in the following four ways:

> Time series – using viewing time directly to emphasize temporal order; to create time sequences and space–time associations. ... Viewpoint transformation – using viewing time to control where and when emphasis is placed (e.g., relocation and zooming in space and time). Re-expression – using viewing time to control what is emphasized, to create attribute sequences and associations (i.e., mapping some variable other than time onto viewing time). Symbolic time – using changes in an object over viewing time as a symbol (e.g., ... with direction of color cycling to show direction of ocean circulation, or rate of cycling to show magnitude of flow) (pp. 228–9)

The interface system should incorporate potential for the user to interact with the animation process itself, tailoring it to meet their needs and to display the results of process modelling. Clearly, it is necessary to develop appropriate principles, guidelines and procedures for the design of dynamic as well as static visualizations for GIS. Such a methodology should utilise a task analysis procedure based on an appropriate model of cognitive interaction.

... CONCLUSION

The role of representations in advanced, interactive information systems is receiving considerable scientific appraisal. This is essential because of the rapidly increasing representational potential of new systems, especially those employing multimedia or hypermedia approaches. In the GIS application domain the potential for systems to produce enhanced graphics (and representations in other modalities) has outstripped the ability of system designers and users to understand which visualizations are most appropriate to particular communication objectives. Visualizations should not be selected merely on the basis of tradition or expediency but rather to provide the greatest utility for users' work requirements. Hence, a rational methodology is required for the design and/or selection of GIS visualizations. The taxonomies suggested here

31

are a first step in developing this methodology to enhance the usefulness of visualization in GIS.

... REFERENCES

Bishop I D, Hull B R (1991) Integrating technologies for visual resource management. *Journal of Environmental Management* 32: 295–312

Buttenfield B P, Ganter J H (1990) Visualization and GIS: What should we see? What might we miss? *Proceedings of the 4th International Symposium on Spatial Data Handling*, Zurich, Switzerland, pp. 307–316

DiBiase D, MacEachren A, Krygier J, Reeves C, Brenner A (1991b) Animated cartographic visualization in earth system science. *Proceedings of the 15th ICA Conference*, Bournemouth, UK, pp. 223–232

Ganter J H (1988) Interactive graphics: linking the human to the model. *Proceedings of GIS/LIS '88*, San Antonio, Texas, pp. 230–239

Goodchild M F (1991a) Integrating GIS and environmental modelling at global scales. *Proceedings of GIS/LIS '91*, Atlanta, Georgia, pp. 117–127

Harrison M D, Thimbleby H W (eds) (1990) *Formal methods in HCI*, Cambridge University Press, Cambridge

Hearnshaw H M (1991) Mental models of spatial databases. *Research Reports* 27, Midlands Regional Research Laboratory, University of Leicester and Loughborough University of Technology, UK

Kraak M J (1991) The cartographic functionality of a three-dimensional GIS. *Proceedings of the 15th ICA Conference*, Bournemouth, UK, pp. 917–921

MacDougall E B (1991) Dynamic statistical visualization of geographic information systems. *Proceedings of GIS/LIS '91*, Atlanta, Georgia, pp. 158–165

Moellering H (1991) Stereoscopic display and manipulation of larger 3-D cartographic objects. *Proceedings of the 15th ICA Conference*, Bournemouth, UK, pp. 122–129

Monmonier M (1989b) Graphic scripts for the sequenced visualization of geographic data. *Proceedings of GIS/LIS '89*, Orlando, Florida, pp. 381–389

Monmonier M (1990) Strategies for the interactive exploration of geographic correlation. *Proceedings of the 4th International Symposium on Spatial Data Handling*, Zurich, Switzerland, pp. 512–521

Moore T J (1991) Application of GIS technology to air toxic risk assessment: meeting the demands of the California Air Toxics 'Hot Spots' Act of 1987. *Proceedings of GIS/LIS '91*, Atlanta, Georgia, pp. 694–714

Olson J R, Olson G M (1990) The growth of cognitive modelling in human–computer interaction since GOMS. *Human–Computer Interaction* 5: 221–265

Rasmussen J, Pejtersen A M, Schmidt K (1990) *Taxonomy for Cognitive Work Analysis*, Cognitive Systems Group, Riso National Laboratory, Roskilde, Denmark

Robertson P K (1991) A methodology for choosing data representations. *IEEE Computer Graphics and Applications* 11(3): 56–67

Sarjakoski T (1990) Digital stereo imagery – integration to geo-information systems. *Proceedings of Commission 3, XVI FIG Congress*, Helsinki, Finland, pp. 489–498

Smart J, Mason M (1990) Assessing the visual impact of development plans. In: Heit M, Shortreid A (eds) *GIS applications in natural resources*, GIS World, Inc., Fort Collins, pp. 295–303

Stephenson T (1990) Imaging, visualization and the challenge of global change: this technology's role in the Mission to Planet Earth. *Advanced Imaging* 5(7): 59–61

Turk A G (1990) Towards an understanding of human–computer interaction aspects of geographic information systems. *Cartography* 19(1): 31–60

Turk A G (1992) *GIS cogency: Cognitive ergonomics in geographic information systems*, Unpublished PhD thesis, University of Melbourne, Australia

van Elzakker C P J M (1991) Map use research and computer-assisted statistical cartography. *Proceedings of the 15th ICA Conference*, Bournemouth, UK, pp. 575–584

Ziegler J, Bullinger H-J (1991) Formal models and techniques in human–computer interaction. In: Shackel B and Richardson S (eds) *Human factors for informatics usability*, Cambridge University Press, Cambridge, pp. 183–206

5 **A TYPOLOGY FOR SCIENTIFIC VISUALIZATION**

K. BRODLIE

...

... INTRODUCTION

The growth of scientific visualization has been one of the most striking features of computing over the past five years. The twin developments of, first, increasing computing power for simulation and, second, increasing band-width of sensing equipment have led to vast increases in data collection in almost all fields of basic and applied science. Visual techniques are an important element both in analysing and presenting these large volumes of data. In geography, there is clearly potential in harnessing scientific visualization systems to help the visual presentation of data from a GIS. In Chapter 6 Gallop gives an overview of the current state of the art in scientific visualization. Modern software systems, such as AVS (Upson et al., 1989) and Iris Explorer (Edwards, 1992), are based on a dataflow paradigm in which the system presents the user with a library of modules. Using a visual programming editor, the user extracts modules from this library and connects them together in a network that will process and display a given dataset in the manner intended.

These systems are extremely flexible and powerful, providing an environment under which the user can manage the visualization process. To support this way of working, it is crucial that there is a sound underlying methodology such as that provided by Haber and McNabb (1990). In the first part of this chapter, we review their work, considering the particular case of interpolating a contour map from a set of scattered data, a common application in geography and other areas. This provides the framework from which to look at various forms of classification in scientific visualization. The aim is to increase the theoretical underpinning of the subject. By appropriate classification of methods and data, one can hope to identify automatically the subset of possible methods for a given set of data. Indeed, with additional information, it may be possible to recommend a particular technique for a particular set of data.

... THE VISUALIZATION PIPELINE

In this section, we draw heavily on the work of Haber and McNabb (1990) who have neatly expressed the visualization pipeline as a sequence of processes. The input to the pipeline is the raw data which may come from a computer simulation (as in Computational Fluid Dynamics) or from measurement (as in remote sensing). Suppose we have topographic data, collected at a set of scattered locations. In almost all cases, these data can be regarded as samples of some underlying entity and, indeed, it is this underlying entity which we wish to display, not the data. For example, in the case study here, we want to see a topographic map of the area, not just the values at the point locations.

The first logical step, therefore, is to construct a model of this underlying entity, based on the data we have collected. If the raw data are assumed to be accurate, this is an interpolation process, fitting a mathematical function through the data points; if the raw data are inaccurate, then an approximation process is required. In both cases, a model emerges. If we suppose that the data are accurate, there is a variety of interpolation methods that can carry out this step: global methods such as kriging and multi-quadrics, or local methods based on triangulation (see Lancaster and Salkauskas, 1986, for example). Haber and McNabb refer to this first step as data filtering or enhancement; we shall refer to it as modelling, to reflect our particular interpretation.

The next logical step is to represent the model geometrically, so that it can be displayed. This is the step where we choose a particular visualization technique, for example, a line graph for 1D data or isosurfacing for volumetric data. In our case the model could be depicted as a contour map, as a surface view, or as an image in which each pixel is coloured according to the value of the model at that point. If we choose a shaded contour map the process involved is the generation of a set of polygonal areas according to heights specified by the user and the output is a geometrical description of the picture to be drawn. It is worth noting here that some visualization techniques choose to display some restriction of the model. In our example, only the transition between areas greater than, and less than, a contour level are shown. Haber and McNabb refer to this step as mapping to an abstract visualization object.

The final step is to render an image from the geometrical description. This will involve the definition of colour tables and other issues specific to the appearance of the picture. In our case, it would involve the preferred choice of colour to show the different areas between contour levels. These different stages of the visualization pipeline are shown in Figure 5.1 which forms the basis for the design of a visualization system. One can provide a set of modelling, mapping and rendering modules; the user can extract an appropriate combination, wire them together into a network and pass data along the pipeline.

Visualization systems, such as AVS and Iris Explorer, largely follow this methodology. However, they tend to blur the modelling operation. A more common pipeline for our case study would be as follows:

(a)   Interpolate on to grid
(b)   Generate contour map from grid
(c)   Render contour map

Step (a) passes only discrete detail of the model, which has to be recreated internally in step (b) and very likely will actually be inconsistent with the model built in step (a).

This section gives us our first example of classification in visualization: we can categorise modules according to whether they perform a modelling, mapping or rendering operation. In the remaining sections, we look at other exam-

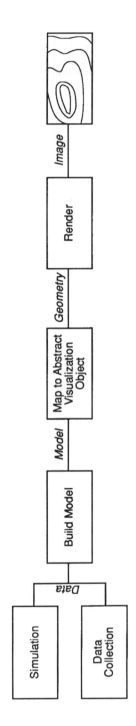

Figure 5.1 The Visualization Pipeline

ples of classification. In the next section, we look at classification of the data that is input to the pipeline; following that, we look at classification of the models which are passed from the modelling to mapping stage. This latter example gives us a classification of visualization techniques, based on the model they are able to depict. The two classifications taken together give a classification of the complete visualization process.

···DATA CLASSIFICATION

In this section, we look at ways of classifying the data which are input into the visualization pipeline.

BERGERON–GRINSTEIN LATTICES

One scheme has been proposed by Bergeron and Grinstein (1989). They introduce the concept of lattices of data. A p-dimensional lattice of q-dimensional data is written $L^p{}_q$. The dimension of the lattice indicates the ordering of the data. Zero-dimensional is unordered (e.g. a set of points), one-dimensional is a vector of data elements (e.g. a list of points to be connected by a polyline, or forming a polygon), and two-dimensional indicates an array of data elements (for example, height and pressure at the nodes of a rectangular grid). The dimension of the data refers to the number of components in a data element. Thus a list of points in 3D is a one-dimensional lattice of three-dimensional data, an object of type $L^1{}_3$. Bergeron and Grinstein go on to describe the visualization process as a sequence of transformations on lattices. The process of interpolating scattered data on to a rectangular grid, for example, would convert a lattice $L^0{}_1$ to a lattice $L^2{}_1$. A failing of their notation is that it fails to reflect the dependent variable involved. Is it, for example, a scalar, a vector, or a number of different components, say, pressure, temperature and velocity as in meteorological applications?

CASE STUDY OF EXISTING SYSTEM: IRIS EXPLORER

Existing visualization systems tend to use this sort of classification. As an example, we take the case of Iris Explorer, recently developed by Silicon Graphics. There are two data types of special interest used by this system. The first is a multidimensional array called a lattice which is very similar to Bergeron and Grinstein's concept. An Explorer lattice is defined by:

(a)   Its dimension, which indicates the connectivity. As in the Bergeron and Grinstein scheme, an image is a two-dimensional lattice, and scattered data form a zero-dimensional lattice.

(b)   The number of elements in each dimension which indicates the number of mesh lines in the lattice (e.g. resolution of image).

37

(c)   The data associated with points on the lattice. Data can be integer, real or string, but must be homogeneous within any particular lattice; there may be several components of data

(d)   The coordinate type, which may be 'bounding box' (regular rectangular grid), 'perimeter' (irregular rectangular grid) or 'curvilinear' (generalised). An entry in the data structure specifies the number of coordinate variables, which essentially means the dimension of the space in which the lattice lies. A two-dimensional curvilinear lattice could, for example, handle a regular grid projected onto the surface of an aircraft wing.

(e)   The coordinate data themselves.

The other data type of interest is the 'pyramid', which supports hierarchical data. It is designed for finite element applications and chemical structures where there is a natural hierarchy implicit in the data. It is represented as a hierarchy of lattices.

... MODEL CLASSIFICATION

As mentioned earlier, the first stage of the visualization pipeline is to create a model of the underlying entity from which the data have been sampled. It is this model which we then 'visualize'. In this section, we look at different ways of classifying this model.

ADVISORY GROUP ON COMPUTER GRAPHICS (AGOCG) SCHEME

A simple classification scheme was developed at a UK workshop on scientific visualization (Brodlie et al., 1992) sponsored by the Advisory Group on Computer Graphics. Some further work has been done on the notation, some of it inspired by the AGI workshop, which produced this volume, and so the treatment here is a modified version of that presented in the above reference. We classify first the type of the model, which can be point, scalar, vector or tensor:

- Point type P. This covers cases where the model is concerned with the existence of an entity at a given point. An example is the presence of a collision between two atomic particles or the existence of a specific geographical feature. Typically, the model simply reflects the input data and the visualization technique used is some form of scatter plot. The data type of the dependent variable is essentially Boolean (yes/no, 0/1). It is useful also to classify the dimension of the space of the independent variables, that is the domain. Thus we classify a 1D Point model as $P(1)$; and in general, an n-dimensional Point model as $P(n)$.

- Scalar type S. This occurs when the dependent variable is a scalar field, for example, the height in a topographic map. Again, we add the dimension, so a topographic map is classified as $S(2)$, a scalar field over a two-dimensional domain. There is an increasing number of volumetric applications where the scalar is defined over three dimensions. For example, in meteorological applications, where the pressure or density is recorded in 3D space above the Earth's surface, the model is classified as $S(3)$.

- Vector type V. Vector fields involve a list of components of the dependent variable, for example, the three components of wind velocity in an atmospheric model. This can be classed as V_m, where m indicates the number of components. Thus, for a 2D vector field over a 3D domain the classification is $V_2(3)$.
- Tensor type T. Tensor fields involve an array of components of the dependent variable. In this case, $T_{3:3}(3)$ indicates a 3×3 tensor field over 3D space.

In the above examples, the domain has been a set of points in a continuous space. There are a few cases, however, where the domain is an enumerated set: for example, the distribution of the particular types of car. We use curly brackets $\{n\}$ to indicate a domain which is an enumerated set in n-dimensional space; thus the distribution of cars of different types is $S(\{1\})$, a scalar field over a one-dimensional enumerated set domain. Similarly there are cases where the model is defined over areas rather than points: for example, the total number of cars in each country. We use $[n]$ to indicate such a domain of n dimensions and $S[2]$ to indicate a scalar over a 2D domain.

This same classification can be used to group visualization techniques, according to the class of model they can depict. Here are some examples:

2D scatter plot: P(2)
Contour plot: S(2)
Surface view: S(2)
2D histogram: S([2])
3D arrows: $V_3(3)$
Isosurface: S(3)

On some occasions, it is useful to display two scalars simultaneously, for example, mean temperature and height over a region. The height is drawn as a surface view, with the surface coloured to represent the temperature value (this is sometimes called a height field plot). This can be coded as 2S(2). Likewise, a meteorological display which shows wind speed and temperature (say by colouring wind arrows according to temperature) would be classed as (S+V)(3).

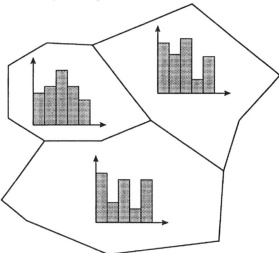

Figure 5.2 An example of S({1})([2])

The domain can also be of mixed type. Consider, for example, a 2D chart showing the number of cars purchased in a particular country, in which one axis is an enumerated set of the different types of car, and the other axis is divided into ranges of months. Such a bar/histogram chart would be classed as S({1},[1]). This notation can also classify some unusual forms of visualization. Consider a chart which shows, for each state in the US, a bar chart indicating the number of people of each ethnic origin; such a chart would be coded as S({1})([2]). The bar chart is S({1}) and it is displayed over areas in a 2D domain (see Figure 5.2). One can almost imagine being able to develop an algebra which would allow operations on the models based on this notation and a table in Brodlie et al. (1992) shows the wide variety of techniques which can be classified using it.

FIBRE BUNDLES AND FIELDS

The above notation does not attempt to take account of the topology of the domain over which the model is defined. For example, both the temperature over a plane surface and the temperature over a curved surface in 3D space such as an aircraft wing (or, indeed, the earth's surface) are coded as S(2). Other authors have taken a more mathematical approach which allows the specification of the range and domain of the model in a precise way. It is simplest to describe their ideas using the example of temperature over an aircraft wing.

Suppose the wing is defined by the surface, S:

$$x = x(s,t); y = y(s,t); z = z(s,t)$$

where s, t are parameters and where (s,t) lie within a region R. Let us denote by F this mapping from the 2D region R to the surface S. Now if the temperature, T, can be expressed as a function, T, of the parameters s,t the temperature over the wing is represented as a 'field', which is a pair of mappings (F,T) which share the same domain R. Some of the early work developing this idea was by Butler and Pendley (1989), who base their description on 'fibre bundles' from differential geometry. Further work, using rather simpler ideas, is reported in Haber, Lucas and Collins (1991).

...CONCLUSIONS AND FURTHER WORK

This chapter has looked at the classification of visualization techniques in a number of ways. First we have seen that modules in a visualization system can be classed as modelling, mapping or rendering according to their function. Second, we have reviewed attempts to classify the data which are input into a visualization pipeline. Finally we have argued that it is important to consider the model of the underlying entity which is being studied, since this is what we really wish to visualize. The notation for classifying models developed in Brodlie et al. (1992) has been improved following discussions at the AGI workshop, and the revised version has been presented in this chapter. It is clear that further improvements to the notation are possible. It would be nice, for example, to develop a corresponding notation for data. This would enable a concise description of a particular visualization pipeline to be developed. The relationship between the underlying model and the display technique also merits further study. In this chapter, we have chosen to classify techniques by the model they represent, suggesting there is a many-to-one relationship between techniques and classes of model. But a visualization technique may choose to represent only a part of the model. For example, in a contour map where constant shading is used between contour levels, the underlying model is S(2) (a 2D scalar field) but the display shows only a restriction of the field, indicating simply the delineation between ranges of values of the field. Should this be encompassed in the notation somehow? Is the 'restriction' a property associated with the technique, or is it an operation performed on the model before the technique is applied? It is evident that more work remains to be done before a solid notation is fully established. The gain however should be significant: a sound structure that will aid the understanding of visualization as a science.

...ACKNOWLEDGEMENTS

Thanks are due to the workshop participants for their helpful comments on this paper. Julian Gallop and Dan Dorling (and earlier Bob Hopgood) made particularly useful observations on the notation, and Hilary Hearnshaw raised searching questions that have led to further improvements.

...REFERENCES

Bergeron R D, Grinstein G G (1989) A reference model for visualization of multidimensional data. *Eurographics '89 Proceedings*, Elsevier Science Publishers BV, pp. 393–399

Brodlie K W, Carpenter L A, Earnshaw R A, Gallop J R, Hubbold R, Mumford A M, Osland C D, Quarendon P (eds) (1992) *Scientific Visualization: Techniques and Applications*, Springer-Verlag, Berlin

Butler D M, Pendley M H (1989) A visualization model based on the mathematics of fiber bundles. *Computers in Physics* 3(5): 45–51

Edwards G (1992) Visualization – the second generation, *Image Processing* 24 (May/June): 48–53

Haber R B, Lucas B, Collins N (1991) A Data Model for Scientific Visualization with Provisions for Regular and Irregular Grids, *Visualization 91 Proceedings*, IEEE Computer Society Press, Washington, pp. 298–305

Haber R B, McNabb D A (1990) Visualization idioms: a conceptual model for scientific visualization systems. In: Nielson G M, Shriver B, Rosenblum L J (eds) *Visualization in Scientific Computing*, pp. 74–93

Lancaster P, Salkauskas K (1986) *Curve and Surface Fitting – An Introduction*, Academic Press, London

Upson C, Faulhaber T Jr, Kamins D, Laidlaw D, Schlegel D, Vroom J, Gurwitz R, van Dam A (1989) The application visualization system: a computational environment for scientific visualization. *IEEE Computer Graphics and Applications* 9(4): 30–42

6 STATE OF THE ART IN VISUALIZATION SOFTWARE

J. GALLOP

... A SIMPLE FRAMEWORK FOR VISUALIZATION SOFTWARE

This chapter is not written from the point of view of a GIS practitioner, but from that of one involved in understanding general purpose software to assess its use in visualization in engineering and science. As might be expected, any dictionary definition of visualization defines it as a human process. Visualization in Scientific Computing (ViSC) is also a human process but it is one which also happens to have modern computer technology associated with it. Data in computers, whether from experimental or simulation sources, are so voluminous and complex that computer assistance, whether through interactive imaging technology or other means, is essential. The goal of ViSC is to provide the tools and methods necessary to provoke human insight into these voluminous and complex data. Perhaps confusingly for the newcomer to this field, when the tools and methods are discussed in the literature the unadorned word 'visualization' is often used to refer to them. Because this has become prevalent, this usage will not be rigorously avoided in this chapter, but will, it is hoped, be obvious. Thus visualization software refers to the computer software used to support the human process of visualization.

Visualization software can be specific to a particular application. For example *Data Visualizer* is appropriate to certain scientific and engineering problems and presents some suitable fluid flow visualization techniques. GIS have similarly restricted domains over which they are appropriate, for example ERDAS, ARCVIEW and GRASS. Other products have emerged as attempts to provide general purpose visualization software. It is the purpose of this chapter briefly to describe the principles behind these general purpose software systems, present a few leading examples, and bring out their benefits and limitations, in a way intended to shed light on whether they will be of use outside science and engineering.

To help us understand how visualization software works, it is useful to subdivide it into simpler computing processes and the flow of data between them. Several authors, including Upson et al. (1989), Haber and McNabb (1990) and Brodlie et al. (1992), have attempted this. A particular contribution by Brodlie et al. (1992) was the emphasis on the possibility of data flowing in two directions, representing control by the human investigator into the computer system in addition to the flow of data resulting in images outwards to the human.

Although the three papers use different terms, superficially it appears that there is a correspondence between the stages each identifies which is shown in Table 6.1. However on closer examination of the operation described by Brodlie et al (1992) as 'building an empirical model', it appears that different authors located it in different places. Building an empirical model is essential to certain methods of contouring scattered data and to model the shape functions in finite element analysis in science and engineering. It is also essential in GIS where examples such as population density estimators may be found. As far as it is possible to judge, the location of the asterisk (*) in Table 6.1 indicates where the paper author(s) located it. Therefore, although the remainder of this chapter uses the Brodlie et al (1992) vocabulary, future work is needed to highlight the role of model building, possibly by identifying it as an operation in its own right.

Table 6.1 Comparison of Frameworks for Visualization Software

	Upson et al. (1989)	Haber and McNabb (1990)	Brodlie et al. (1992)
Stage 1	Filter	Data (*) enrichment	Data manipulation
Stage 2	Mapping (*)	Visualization Mapping	Visualization Technique (*)
Stage 3	Render	Render	Base graphics system

For our purpose, a simplified version of the Brodlie et al. (1992) framework is sufficient and is shown in Figure 6.1. This simple framework distinguishes the role of the visualization technique from the base graphics system. It also distinguishes the different meanings attached to different data streams since

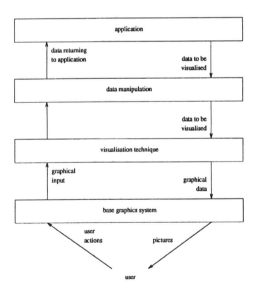

Figure 6.1 Simplified frameword for visualization software

data to be visualized are distinct from graphics data. The framework also shows the possibility of modules communicating data in either direction, allowing user input to be dealt with and passed upstream to the application if desired. In addition, a visualization system should include the possibility of manipulating the data in preparation for delivering them to a visualization technique so this is distinguished as a separate process. An example is the selection of one variable from a set of multivariate data (data manipulation); if it is in three-dimensional space, an isosurface could be generated (visualization technique); and the resulting polygons might then be rendered (base graphics system).

... VISIBILITY OF THE FRAMEWORK: A CLASSIFICATION

Some visualization systems such as AVS, make this framework visible to the user. AVS allows the user to select filters, mappers and renderers (see Table 6.1 and Upson et al. 1989) and connect them together using a graphical network editor. In AVS, the flow of data is explicitly defined by means of connecting pipework between modules. A system with this characteristic is referred to as a data flow visualization system or alternatively an application builder visualization system. In the academic community in the UK, an evaluation of application builders has recently been published by the Advisory Group on Computer Graphics (AGOCG, 1992).

AVS (AVS Inc), apE (formerly from Ohio State University but now from Tara Visual Inc.), Iris Explorer (Silicon Graphics), Khoros (University of New Mexico) and Visualization Data Explorer (VDE from IBM) share the following characteristics of application builders:

- The user can provide new data formats. If the data model of the user's data is shared by the visualization system.
- The user can connect modules by visual programming methods.
- The user can create and supply modules.
- Although they have made advances in the visualization of data defined in 3D space, compared to more traditional products such as computer cartography systems, they are somewhat primitive in their solutions for data defined in 2D space.

Other visualization systems conceal the framework from the user. Their strategy is to provide a fixed set of techniques with a tailor-made user interface. Examples of such a system include Data Visualizer (from Wavefront Technologies Inc.) and the Uniras interactives (from Uniras). Typically such systems are referred to as turnkey systems. Some turnkey systems (for example Data Visualizer) do allow data formats to be specified by the user, but only if the data model is acceptable to the system. If a turnkey visualization software system contains the solutions that the user needs, this will probably be the preferred approach. The integrated nature of a turnkey system can have advantages of speed and reduced memory usage whereas application builders are liable to suffer problems with memory management in that, because of the separation into modules, temporary data storage is not released back to the system.

The fact that the application builders are configurable gives them a number of advantages:

- Visualization problems are diverse: if the user's problem falls outside the provided set, a solution can be added by that user.
- The user community can add modules and this effectively increases the 'provided set', before one needs to write one's own modules.
- They are usually general purpose and allow methods and modules to be exchanged between disciplines.

Are the application builders general purpose enough to cover GIS requirements? Is there a turnkey system which precisely suits GIS purposes?

... DATA MODELS IN VISUALIZATION

Data to be visualized comes in a wide variety of forms that at first sight appear to be impossible to deal with in any coherent way. However, if we put aside the matter of the format itself and concentrate on the data model, there are some common characteristics that we can identify provided we:

- Distinguish between the data values, that is, the dependent variable, space, and the space on which the data values are defined, that is, the independent variable space. Thus, barrels of oil exported (dependent) may be defined for a set of countries (independent) and the temperature and flow velocity of a fluid (dependent) at a set of points in 3D space (independent) can be specified.
- Take care with dimensions. When stating the dimensions of the problem, these two spaces are often confusingly rolled together: thus 'I have a 5D problem' or 'my system handles 4D data'. In fact 1D data in 3D space are very different from 3D data in 1D space. In our examples, the barrels of oil exported are represented by 1 scalar at each point in 1D space; the temperature and flow velocity problem is represented by 1 scalar and a 3D vector at each point in 3D space.
- Specify the geometry of the independent variable space for some problems. Examples are polar or spherical coordinates or (u, v) over a piecewise curved surface.
- Note that the connectivity of independent variable space is inherent in finite element problems, where the solution data may be defined, for example, at the centroids of irregular polygons or polyhedra that constitute the mesh.
- Take advantage of any regularity in the data. If, because of this regularity some information can be held implicitly, this reduces the quantity of data stored and in many cases makes the algorithms more straightforward and efficient. Some visualization systems require the sample points in the independent variable space to be regular. Khoros is an example of this. Other visualization systems also accept unstructured data, databased on a mesh of polygons (or polyhedra) with no inherent structure, for example, AVS, Iris Explorer, VDE and Wavefront's Data Visualizer.

As the number of dimensions is increased (both in the dependent and the independent variable spaces), so the problem becomes more difficult to visualize. Most general purpose visualization systems have the 'hooks' that allow data of higher dimensions to be imported, but these are as far as most available systems go in this respect.

45

What data models do GIS users expect to be supported by visualization systems? One area of weakness, as I see it, is that visualization systems are not set up explicitly to generate maps. Are the data models sufficiently powerful that this is not a problem?

... VISUALIZATION TECHNIQUES

Having understood the data, it is possible to select a visualization technique from those available (see Chapter 5). For example, data consisting of 1 scalar defined in terms of one independent variable can be represented by a graph. Temperature and flow velocity in 3D space can be represented by a set of 3D arrows whose colour is controlled by the temperature. However, for higher dimensional spaces there is a lack of visualization techniques supporting higher dimensional data, even though methods have been described in the literature.

The techniques provided with most visualization systems do give some insight that was not previously possible. With most users though, one has the uneasy impression that they are not in control of the process, that the mapping from their data to the geometry is not explained.

The important point about the application builders is that the investigator can replace predefined visualization techniques by ones more suitable for the investigation in hand. What are the requirements of GIS users and are they satisfied by the techniques provided by present day visualization systems?

... CONTROL

There is a number of characteristics of how users control visualization systems that are important.

- Distribution. The user may need to use remote computing equipment because insufficient computing power or data storage is available locally. Most application builders allow modules to be remotely executed.
- Multiple views. The phenomena and data that are being visualized are often sufficiently complex that the user needs more than one way of viewing the data simultaneously.
- Interactive control. It is important to know whether or not the investigation has interactive control of the application when using a visualization system. Three sorts of interaction have been described. In post-processing the application generates a results file and the visualization software allows the investigator to display it. In tracking the investigator may view the results while the application is running, and, in steering the investigator not only views the results while the application is running but also may direct the process.
- Post-process. The simplest approach for the developer is for there to be a loosely coupled approach between the application and the visualization system. In this way the application and the visualization system communicate by files and their user interfaces are quite separate from each other. This corresponds to the traditional post-processor approach.
- Inclusion. Simple application modules can be included within the visualiza-

tion system. In principle all the application builder visualization systems allow this.

- Integration. Inserting application modules may be difficult if the application has its own user interface and database access methods. For example, in a GIS, the user may interactively select a region of the world, select the type of information to be worked on and select the style of visualization. This may only be possible if some tightly coupled connection of the application and visualization has been developed. This is difficult for most visualization systems unless a completely modular approach has been adopted.

Again, what are the requirements of GIS users in this regard?

... SUMMARY

A number of characteristics of visualization software have been outlined dealing with the visibility of the framework, data models, techniques and control. The benefits and limitations of general purpose visualization software have also been described.

... REFERENCES

Advisory Group on Computer Graphics (AGOCG) (1992) Evaluation of Visualization Systems. *Advisory Group On Computer Graphics Technical Reports*, 9

Brodlie K W, Carpenter L A, Earnshaw R A, Gallop J R, Hubbold R J, Mumford A M, Osland C D, Quarendon P (eds) (1992) *Scientific Visualization: Techniques and Applications*, Springer-Verlag, Berlin

Haber R B, McNabb D A (1990) Visualization idioms: a conceptual model for scientific visualization systems. In: Nielson G M, Shriver B, Rosenblum L J (eds) *Visualization in Scientific Computing*, pp. 74–93

Upson C, Faulhaber T Jr, Kamins D, Laidlaw D, Schlegel D, Vroom J, Gurwitz R, Van Dam A (1989) The application visualization system: a computational environment for scientific visualization, *IEEE Computer Graphics and Applications* 9(4): 30–42

SECTION B

ADVANCES IN VISUALIZING
SPATIAL DATA

7 Introduction to Advances in Visualizing Spatial Data

A. MacEachren, I. Bishop, J. Dykes, D. Dorling & A. Gatrell

... Visualization Techniques for Spatial Data

There are many approaches to visualizing spatial data. There is variety in the intended purpose, proposed audience, level of interactivity and degree of intimacy with the data themselves. The chapters in this section illustrate this variety together with some of the ways modern visualization ideas are helping to solve some very old cartographic problems. Clearly the work represented here is not the sum total of advances in visualization of spatial data, nor is it intended to be, but we hope that it will give readers a reasonable overview as well as pointers to some of the work that has not been included.

In order to understand how the chapters fit within the broad framework of visualization approaches it is appropriate to set the context and provide some definitions. Our basic organisational tool in this introduction is a simple typology of visualization techniques for spatial data based on distinctions drawn according to four dimensions of variability. The first distinction to be made is in *purpose*. DiBiase (1990) proposed the model of how visualization tools are used by science shown in Figure 7.1. This was derived from Tukey's sequence of exploratory, confirmatory and presentational statistics, and suggests that visualization tools are used by science for a range from data exploration and confirmation (prompts for visual thinking) to the synthesis and presentation of ideas (vehicles for communication). We will refer to the ends of this continuum as communication and ideation and these form end-points on the first dimension of our typology.

Scientific visualization has been developed primarily to help scientists extract ideas from masses of multidimensional data. As the chapters in Section A have argued, the emphasis on ideation is at the exploratory end of this purpose continuum. Such visual data exploration techniques have not yet been heavily used in GIS but are being applied elsewhere in, for example, seismology, climatology and meteorology (see DiBiase, 1990, for an overview). In contrast to the general emphasis in scientific visualization on exploratory analysis and ideation, much of the work linking visualization to GIS done by the landscape architects is directed towards communication in which designers use visualiza-

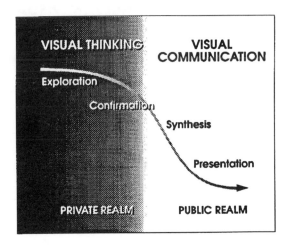

Figure 7.1 DiBiase's model of the range of uses to which maps can be put in geographical inquiry

tion tools to present plans to clients or decision-makers. Running approximately parallel to this distinction in purpose is one in audience. As suggested above, the principal consumer of visualization tools supporting ideation is the scientist/data analyst. The audience for communicative visualization is the non-expert public. In the latter case the audience may be large numbers whereas in the former it is frequently a single individual.

A second dimension in our multidimensional visualization application space is the *degree of interactivity available* to the audience shown in Figure 7.2. While an archive of static images or movie clips can be appropriate for communication to the public, the scientist interested in posing 'what if?' scenarios will generally demand an interactive capability to get the most insight from the data. As technology allows, interactive GIS based visualization will be taken increasingly into the public arena in order to permit, for example, a public meeting to debate the

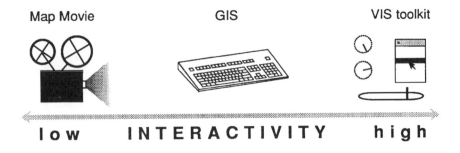

Figure 7.2 The range of interactivity available with various visualization tools

consequences of alternative resource management strategies. The need for interaction in presentational visualization tools will, therefore, increase.

A third axis in this non-orthogonal set of dimensions is the *degree of abstraction* in the visualization. As Figure 7.3 shows, this can range from the highly symbolic to the highly detailed and realistic. Abstract representations may be largely unfamiliar to someone not trained in visualization tools use, while less abstract representations can provide a view that is both familiar and accessible

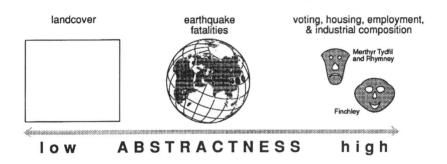

Figure 7.3 The abstractness range of visualization tools

to workers in other disciplines or to the general public. This continuum, therefore, to some degree parallels the audience axis. Depictions at the realistic end of the continuum facilitate representations of the physical world (whether landscape or human anatomy) and identification by the viewer of what occurs at particular locations. Depictions at the other, abstract, end facilitate the contemplation of abstract concepts and may even use space to depict things other than space such as a multidimensional scaling solution showing the position of GIS in geography's disciplinary 'space'.

Finally, spatial visualization is distinguishable according to the *aspect of the phenomenon to information process* about which is it concerned. One end of this range involves concern about how well the data representation available in a GIS matches the phenomenon for which it provides a model and/or how to restructure and combine information available in a GIS to produce an appropriate model of the spatial information to be visualized. The chapters in this section by Gatrell, Bracken and Dorling emphasise these issues. Included at this level are considerations of data about data (metadata) dealing with accuracy, classification and so on. Complementary to this focus on how data represent phenomena is a concern about how displays represent data. Issues of symbolisation choice and generation are relevant here. In this section Bishop explores the use of techniques for the photo-realistic presentation of spatial data, Dorling discusses the use of Chernoff faces as multidimensional symbols on a schematic cartogram base, and MacEachren and Openshaw, Waugh and Cross focus on the graphic building blocks or variables that become available when we move from a static to a dynamic visualization environment.

...WHERE ADVANCES ARE BEING MADE

Figure 7.4 attempts to show where the work described in this section falls in relation to three of the dimensions identified above. With suitable visualization tools we could map each contribution into the complete multidimensional space but as yet this would be hard to convey to the reader on the printed page. Instead each axis is shown independently, with the span of each author's activity shown by an arrow whose head indicates the direction in which they anticipate their work will move as technology allows. Four trends emerge from these figures which are worthy of discussion.

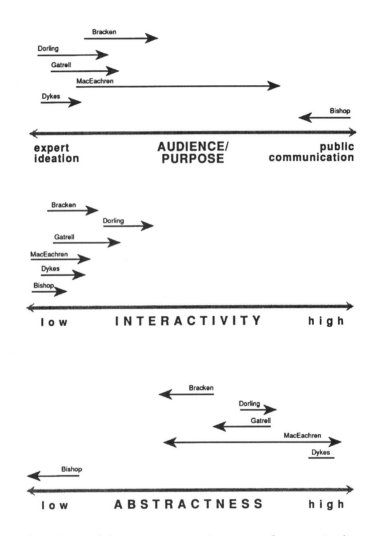

Figure 7.4 Emphasis of chapters in Section B in terms of purpose/audience, interactivity and abstractness of display

AUDIENCE

Most of the applications described match the overall trend in scientific visualization to emphasise tools for use by experts in scientific exploration. With his emphasis on presentational tools accessible to non-experts in a public policy context, Bishop is the main exception. There is general agreement that, for visualization to benefit non-scientific applications of GIS, it will be necessary to direct more attention to visualization materials suited to a wider audience. In this regard, we are seeking to improve our understanding of rendering such that it becomes familiar and accessible to the public. This may also involve some degree of public education to introduce the new symbolism of abstract visualization. The typical GIS emphasis on applied over basic research and public presentation over individual scientists manipulating a dataset offers our greatest challenge in adapting principles of scientific visualization for use with GIS.

Although not apparent from these contributions, the midpoint of the continuum between private exploratory analysis and public presentation has not been ignored by those interested in geographic visualization tools. One of the most significant contributions is Monmonier's (1992) Atlas Touring project. In this, Monmonier has developed a hypermedia environment that guides users through a network of information using a combination of maps, graphics and text. The system is based on his concept of graphic scripts as structured 'narratives' for presenting spatial information. One of Monmonier's goals, with implications across the whole exploratory to presentational continuum, is to automate the production of graphic phrases (subsections of overall graphic scripts) that act to focus a user's attention on 'interesting' views selected by the system.

ABSTRACTION

Most of the contributions to this section emphasise highly abstract representations of spatial phenomena. Several contributors, however, are seeking to reduce the abstract nature of the rendered product as a step toward widening audiences. At the same time, others point out that the advantages of abstracting reality, as demonstrated by centuries of cartographic success, will remain important in the context of expert analysis as a means of coping with increasingly complex and voluminous datasets. At a level of abstraction beyond that emphasised by the contributors to this section, significant advances have been made by Cleveland and his colleagues (Cleveland and McGill, 1988) in Exploratory Data Analysis (EDA). EDA researchers have made innovative use of tools developed by others for realistic rendering, such as stereo views, shading and other depth cues, as a means for depicting complex data relationships in abstract data spaces. In the context of land cover analysis from remotely sensed images, Hodgson and Plews (1989) adapted these techniques to design of an n-dimensional extension of the commonly used co-spectral plot. Cleveland and others have also given considerable attention to design of sophisticated interactive data manipulation tools. Monmonier (1989) has adapted one of their techniques ('scatterplot' brushing) for application in a spatial data analysis ('geographic' brushing) and DiBiase et al. (1992) have extended the method to a temporal context. In addition to Bishop's contribution, there is also a variety of other efforts underway to take advantage of realistic rendering tools in GIS visualizations. Among the most notable is Moellering's (1989) use of stereo display technology that allows an image on a single monitor to appear three-dimensional, and the application of three-dimensional rendering and image filtering techniques to visualization of plant growth and pest models by Orland et al. (1992).

INTERACTIVITY

All contributors to this section see a need for increasing the interactivity of visualization tools linked to GIS. Achieving this depends, however, on hardware and software developments in the GIS environment particularly the derivation of appropriate data models and displays for dynamic analysis. Although all the developments presented here have implications for real-time interaction with spatial data embedded in a GIS, only limited forms of interaction have yet been explored. The interactive exploration of GIS datasets will be as critical to visualization in GIS as it has been for ViSC in general. If appropriate data models and display forms are not developed, however, this interaction is as likely to mislead as to provide insight.

Although in this section we chose to emphasise the phenomenon-to-data model and data model-to-display stages of visualization environments, there have been several researchers who have addressed the issue of interactivity in spatial data analysis. Monmonier's (1989) geographic brushing mentioned above was among the first examples but perhaps the most explicitly GIS oriented application of interactive visualization tools is MacDougall's (1992) *Polygon Explorer*. This prototype software is specifically directed at statistical visualization of polygon data, which MacDougall argues is the most common data structure used in GIS. *Polygon Explorer* links statistical graphics of a variety of types to a polygon database and allows the analyst interactively to probe and query in both spatial and aspatial ways by pointing to locations on maps or graphs. Tools for identification of outliers and for regional grouping by cluster analysis are also included.

Interactive visualization of spatial data, if it is to go beyond the experimental stage of the *Polygon Explorer*, will require highly efficient data storage and extraction. This graphic and attribute data management function is best performed by GIS. Most of the applications described in this section have involved the use of GIS somewhere in preparation of their visual displays but in no case has this been a seamless operation. Each researcher has had to write transfer programs and tie their operations to the GIS with 'boot laces and sticky-tape'. Functional interactivity will require something better than this. Not only will GIS vendors have to incorporate additional visualization options in their tool kit, they will also need to speed up data extraction, processing and screen-draw operations, taking advantage of such developments as object oriented data management and the specialised graphics processors offered by visualization oriented workstations. It is unlikely that we will see sophisticated visualization tools incorporated directly in GIS software in the near future. What we must work towards is smoother links between GIS and visualization systems, something that will require vendors to become more sensitive to the role of visual display as more than a final output and that will require visualization system vendors to develop ways in which their tools can be linked directly to the complex space–time data structures required in GIS.

THE PROMISE OF DEVELOPING TECHNOLOGY

A variety of emerging computer technologies have the potential to contribute to the visualization of spatial data. Interactive stereographic displays are now available and should prove valuable to the individual scientist seeking insight into three-dimensional data. Movies for communication can also be projected in stereo for wider audiences. In both movies and at workstations, sound will become more than the overlay of narration to accompany visualization events. As an information presentation variable, sound is being actively pursued by a number of researchers. In addition, multimedia tools increase the possibility to link visual displays, sonic feedback, graphics and text in ways that give users of GIS-visualization environments multiple perspectives on problems. Beyond multimedia, virtual reality systems may offer additional possibilities for extending visualization toward realism. Currently, virtual reality developments involving head mounted displays and body gloves are cumbersome, of low graphic quality and expensive. While the idea of walking/driving/skiing through a simulated landscape is clearly attractive, it is less clear whether or not the ability to take a virtual stroll through an abstract 3D graphic such as Dorling's (1992) three-dimensional space–time cartograms is of any real advantage.

...INTRODUCTION TO ORDER OF PAPERS

Having provided our typology of techniques, and stipulated that the axes on which we locate our work are neither mutually independent nor ranked in terms of significance, it is evident that defining a vector through four dimensional, non-orthogonal, research space by which the following chapters might have been ordered was a complex task. Several groupings occur within the section, and it is stressed that for maximum gain the section should be read as a whole. According to their interests and perspectives, individual readers may find the continua illustrated in Figure 7.4 of use in ordering their reading of the section. Although several orders seemed suitable, the section has been ordered roughly by the degree of correspondence between the visual product discussed in each chapter and reality.

At the realistic end of this continuum, Ian Bishop advocates the use of visual familiarity as a tool for effectively communicating spatial process. He argues that realistic scenes can be used at either end of the audience continuum to convey multiple scenarios to a non-expert audience, or to facilitate the analysis of model success to the scientist. There follow two papers dealing with the zone-based enumeration data common in much geographic work. These provide examples of ways in which the traditional choropleth map may be superseded if additional locational data are integrated with the area-valued data. Tony Gatrell outlines the restrictions of this data type and shows how zone centroids can be used to estimate a spatially continuous density surface. Through this transformation the visual product bears a closer resemblance to the surface from which the data are collected than does the traditional choropleth map. A similar transformation is illustrated by Ian Bracken who focuses on using zone-based data to create a raster model using a weighted decay function as an approximation to a spatially continuous surface of population density. The results are visually appealing, but, just as important, the data structure produced is compatible with the majority of GIS.

Less familiar, more abstract representations are discussed in the next two chapters. These also deal with zone-based enumerated data, but focus on the visualization of topological, rather than metric, relationships. A set of transformations are discussed by Dan Dorling, who puts the case for using cartometric space as a necessary part of analysing the spatial nature of social structure. He proposes severing the strong traditional link between land area and mapped area in favour of a particular (non-spatial) social variable as the spatial metric. A variety of techniques can be applied to the resulting spatial arrangement, including the surface generators described above. Jason Dykes continues the theme of analysing relationships in non-geometric space showing how he uses visual representations of these relationships to improve more traditional displays.

Whereas Chapters 8 to 12 focus on ways of improving traditional representations, those by Alan MacEachren and by Stan Openshaw, David Waugh and Anna Cross outline the variables available to the visualizer when maps are animated. Whether viewed as a movie or by interactive browsing, the ability to order in sequence and step through a set of scenes adds to the communicative or investigative cartographic bandwidth. The methods described can be applied to any of the preceding techniques, for example a time series of socioeconomic rasters or bulging population surfaces illustrating social change, Dorling's swinging vote cartograms and Dykes' co-occurrence volumes with fly-by and dynamic iso-surface variation, or real time update from base-map modification (re-expression). It seems appropriate to conclude this introduction by this indication of the wide variety of combinations of spatial and temporal cartographic variables that are becoming available to visualizers as technology improves.

... REFERENCES

Cleveland W S, McGill M E (1988) *Dynamic Graphics for Statistics*, Wadsworth & Brooks/Cole Advanced Books & Software, Belmont, Calif.

DiBiase D (1990) Visualization in the earth sciences. *Earth and Mineral Sciences* Bulletin of the College of Earth and Mineral Sciences, The Pennsylvania State University 59(2): 13–18. Requests for a copy of this paper should be sent via e-mail to: dibiase@essc.psu.edu

DiBiase D, MacEachren A, Krygier J, Reeves C (1992) Animation and the role of Map design in Scientific visualization. *Cartography and GIS* 19(4): 201–214

Dorling D (1992) Stretching space and splicing time: from cartographic animation to interactive visualization. *Cartography and GIS* 19(4): 215–227

Hodgson M E, Plews R W (1989) N-dimensional display of cluster means in feature space. *Photogrammetric Engineering and Remote Sensing* 55(5): 613–619

MacDougall E B (1992) Exploratory analysis, dynamic statistical visualization, and geographic information systems. *Cartography and GIS* 19(4): 237–246

MacEachren A M, Buttenfield B, Campbell J, DiBiase D, Monmonier M (1992) Visualization. In: Abler R, Marcus M, Olson J (eds) *Geography's Inner Worlds*, Rutgers University Press, New Brunswick, NJ pp. 99–137

Moellering H (1989) A practical and efficient approach to the stereoscopic display and manipulation of cartographic objects. *Auto-Carto 9, Proceedings, Ninth International Symposium on Computer-Assisted Cartography*, Baltimore, Maryland, pp. 1–4

Monmonier M (1989a) Geographic brushing: enhancing exploratory analysis of the scatterplot matrix. *Geographical Analysis* 21: 81–84

Monmonier M (1992) Authoring graphic scripts: experiences and principles *Cartography and GIS* 19(4): 247–260

Orland B, Onstad D, Obermark J, LaFontaine J (1992) Visualization of plant growth and pest models. *Technical Papers*, ASPRS/ACSM/RT 92, 5, Resource Technology, 246–253

8 THE ROLE OF VISUAL REALISM IN COMMUNICATING AND UNDERSTANDING SPATIAL CHANGE AND PROCESS

I. BISHOP

... INTRODUCTION

It is often argued that GIS are a device for turning data into information. One of the key aspects of this transition from data to information is effective communication, first between the GIS and the user and, second between the user and other information consumers. These consumers are scientists, decision-makers and the interested public. Each has different communication requirements, according to their knowledge base. A range of visualization procedures is necessary to respond to various needs.

The scope of the GIS based data which we seek to visualize and communicate is also varied and changing. GIS began life as essentially two-dimensional systems. They stored, manipulated and produced maps. They are now progressing towards three dimensions as the CAD/GIS gap is being gradually closed and we have begun to look towards full 4D status in our spatial data handling systems. At the same time GIS are being asked to deal with data at a variety of scales, resolutions, details and levels of abstraction from the global resources database to the developer's site plan. In terms of the broad applications areas of GIS, effective display of 4D information in many guises is required for responding to both planned and unplanned changes in our environment.

Visualization technology has to date been largely concerned with communication, to the scientist, of scientific data:

> Scientific breakthroughs depend on insight. In our collective experience, better visualization of a problem leads to a better understanding of the underlying science, and often to an appreciation of something profoundly new and unexpected. (McCormick et al., 1987)

Yet the public concerned with spatial change and process has different requirements:

> For centuries it appears to have been assumed that a drawing-is a drawing-is a drawing, and that it probably means the same thing to all who view it. The evidence is sparse and scattered but it does suggest that this assumption is invalid. The limited evidence also suggests that the most realistic simulations, those that have the greatest

similitude with the landscapes they represent, provide the most valid and reliable responses. (Zube et al., 1987)

At work here (in a high-resolution oblique aerial photograph) is a critical and effective principle of information design. Panorama, vista and prospect deliver to viewers the freedom of choice that derives from an overview, a capacity to compare and sort through detail. And that micro-information, like smaller texture in landscape perception, provides a credible refuge where the pace of visualization is condensed, slowed, and personalised. (Tufte, 1990)

The non-scientific audience, in other words, wants abstraction minimised, information content maximised, and with the whole package digestible and non-threatening. This suggests the use of a visual realism approach which shows information consumers what will or might happen under a variety of conditions and permits them to explore the alternative environments using their natural sensory perceptions. Those not trained in interpretation of abstract information should be able to work at this more intuitive level. As decision-making becomes increasingly an exercise in public consultation and compromise, decision support requires that all aspects of a project be clearly understood by the public.

...THE BENEFITS OF REALISM

A visualization technique which transcends skill, experience and awareness levels also reduces the scope for differences of opinion brought on by differences in interpretation. The different backgrounds of individual information consumers give great scope for differential interpretation of a common set of visualization stimuli. As Petch in Chapter 23 points out 'acquisition of spatial information is selective or partial. It is generally oriented and only accessible in a certain way.' He goes on to state that our recognition of patterns, global structures or hierarchies occurs 'in our minds and they develop in our minds in all sorts of different ways' and that 'spatial information gained from direct experience is more orientation free and useable'. Realistic simulation is the nearest we can achieve to direct experience and has been referred to as the 'natural scene paradigm' of visualization (Robertson, 1991). Direct experience is, in any case, not available when alternative futures are being assessed.

In Chapter 21, Hearnshaw points out that visualization frequently involves 3D perception from 2D displays and that this can also create ambiguity in interpretation. With the normal options of relative parallax and changing focus not available to aid us, any clue which helps to reduce the ambiguity of distance is a benefit to the visualization consumer. In this context realism provides the great advantage of familiarity to aid interpretation: we know the true size relativities of objects in the field of view. This advantage can be further enhanced by the use of haze, simulated depth of field and the natural movement of objects within the visual field.

The tendency of humans to ignore stimuli in an image if they contradict attempts to organise the image into a real world scene has been identified by both Haber and Wilkinson (1982) and Robertson (1991). Once again, therefore, the opportunity for misinterpretation is limited if the image is already a real world scene. This is not to argue that any data presentation exercise is amenable to realistic presentation. However, if the option exists the advantages should be exploited.

If visualization is to extend beyond the communication role and so lead to scientific insights not otherwise available, then it must be well tuned to human processes of creative thinking. According to an increasing number of theorists effective creative thinking relies on two aspects of our thought processes. These are referred to (Edwards, 1989) as right-brained-mode (concrete, spatial, intuitive and holistic) and left-brained-mode (symbolic, digital, logical and linear). Given this bifurcation in thought process and the evidence that both modes have important roles, it is appropriate to have available information stimuli which appeal to both ways of thinking. We should combine the symbolic with the concrete and realistic.

... APPLICATIONS OF REALISM

In principle, realistic visualization can be applied in any situation in which a change in the visual environment is anticipated. There is, however, a danger in the use of realistic visual simulation as a medium of communication and particularly in regard to natural resources planning and management. The concern is that aesthetic aspects may come to dominate other possibly more important variables such as those related to ecology, hazards, costs, soil loss and so on. While this may be true in terms of the traditional use of visual simulation to portray 'before and after' static representations of new power stations or quarry sites, new trends in visualization indicate a much broader range of purposes.

Because the technology required for visually realistic presentations of alternate realities is generally also capable of supporting animation the following discussion assumes animation as an automatically available option in application of the natural scene paradigm. That is, realism supports the visualization of change. Change has been categorised by DiBiase et al. (1992) as change in either: (a) the spatial location of the observer (flybys); (b) the spatial location and/or attributes of objects (time series); or (c) position of objects within attribute space (re-expression). Change of the first two categories is clearly amenable to realistic presentation.

Visual realism, using changes in viewer position or of objects and attributes, may prove the best approach to public understanding of many consequences of change beyond the aesthetic. A simulation of projected traffic levels and the driving experience associated with those levels is clearly more intuitive, but no less information rich, than model data on vehicles per hour. Simulated changes in noise levels (auralisation?) in the main street kindergarten mean more to parents and teachers than predicted decibel readings. Water lapping at specific door-steps will be more compelling than a redrafting of the 100-year flood plain boundary in communicating to local councillors the impact of up-catchment urbanisation. A 'virtual' fly-over of Madagascar as it will be in thirty years given recent forest-clearing practice may communicate the issues better than a mass of statistics.

Of course, the detailed planning and management of change still require that planners and managers have access to vehicle projections, dB(A) estimates, flood plain boundaries and species distribution statistics, but the visualization process is the key to informed public evaluation of their activities. Scientists, modellers and data analysts also stand to benefit from the availability of realism in visualization. Fire behaviour modelling, for example, has been the subject of considerable work in Australia and elsewhere. When tested against historical fire data some models are found not to perform particularly well on some fires. Explanations of this failure are often hard to find. If the models were used to drive realistic simulations the reasons for divergence between models and reality may very quickly become apparent to experienced fire-fighters who are unable to respond to abstract representations.

... Enabling technologies

There are on-going technological trends which will permit visualization to become more fine-grained in both spatial and temporal terms and thus better able to deal with the diversity of data stored within GIS and to meet the needs of a wider audience. These include:

1 More dynamic modelling tools within the GIS environment. Commercial vendors are reintroducing raster capability into their systems while researchers are developing dynamic modelling command languages and tool kits which will eventually find their way into commercial products (Green et al., 1989).
2 The use of photographs or video (both aerial and ground-based) as texture maps to increase the realism of GIS generated perspectives is becoming well established (Kaneda et al., 1989; Bishop and Flaherty, 1991). This is illustrated in Plates 1 and 2. In one case the textures used are scanned aerial photographs from the area being modelled. In the other, generic textures from a library of representative Australian rural aerial imagery are used. These images were scanned using a PC based flatbed scanner and transferred to a Macintosh for colour and contrast adjustment using commercial software. They were then transferred to a UNIX Workstation where Wavefront Technology's Advanced Visualizer software was used to convert the images to textures. The terrain models were converted from an Intergraph-generated TIN model in one case and from a raster height field in MAC-GIS in the other. Bespoke software was used to adjust the height field in forested areas and add edge polygons.
3 Animators have largely solved the problems of realistic representation of fluid elements such as water, fire or a field of grass in three-dimensional motion (Reeves and Blau, 1985).
4 Although present computer processing power does not usually make real time computation of this imagery possible to a satisfactory degree of realism, computer controllable, high capacity video disks allow a large number of sequences representing different scenarios to be stored and explored interactively as highly realistic imagery (Bishop, 1992).
5 Display options are expanding. Although the individual tactile interaction available through head-up displays and data gloves cannot yet be accommodated, we have barely begun to explore the potential for extending the realism of environmental display through stereo monitors, stereo projection or panoramic projection devices.

6 The loss of immediate access to numerical values (available through colour-coded symbolic visualization) could be overcome using developments in 'virtual probing' to read numerical values off simulated objects identified using an interactive probe in 3D space.

... CONCLUSION

While the technological frontiers continue to push forward, ever expanding our windows of opportunity, we should not accept the opportunities provided uncritically. Dangers remain and need researching. The ultimate virtual reality (the next generation Holodeck) is still many years and many currency units away. All our efforts to date remain abstractions and as such open to misinterpretation. We need to test, and retest, the validity of any visualization process as an accurate means of communication of change and process. We need also to determine what level of abstraction is valid and acceptable in what set of circumstances so as to avoid unnecessary escalation of costs and expectations among information consumers.

... REFERENCES

Bishop I D (1992) Data integration for visualization: the role of realism in public debate. *Proceedings AURISA'92, Australian Urban and Regional Information Systems Association*, Gold Coast, Australia

Bishop I D, Flaherty E (1991) Using video imagery as texture maps for model driven visual simulation. *Proceedings Resource Technology '90*, Washington, DC, pp. 58–67

DiBiase D, MacEachren A, Krygier J, Reeves C (1992) Animation, abstraction and the role of cartography in scientific visualization. *Cartography and GIS* 19(4): 201–214

Edwards B (1989) *Drawing on the Right Side of the Brain*, Tarcher, Los Angeles

Green D G et al. (1989) A generic approach to landscape modelling. *Proceedings Simulation Society of Australia Conference*, Canberra

Haber R N, Wilkinson L (1982) Perceptual components of computer displays. *IEEE Computer Graphics and Applications*, May, pp. 23–35

Kaneda K, Kato F, Nakame E, Nishita T, Tanaka H, Noguchi T (1989) Three-dimensional terrain modeling and display for environmental assessment. *Proceedings ACM Siggraph '89*, Boston

McCormick B H, DeFanti T A, Brown M D (eds) (1987) Visualization in scientific computing. Special issue ACM SIGGRAPH *Computer Graphics* 21(6)

Reeves W T, Blau R (1985) Approximate and probabilistic algorithms for shading and rendering structured particle systems. *Proceedings of SIGGRAPH '85*, pp. 313–322

Robertson P K (1991) A methodology for choosing data representations. *IEEE Computer Graphics and Applications* 11(3): 56–67

Tufte E R (1990) *Envisioning Information*, Graphics Press, Cheshire, Connecticut

Zube E H, Simcox D E, Law C S (1987) Perceptual landscape simulations history and prospect. *Landscape Journal* 6: 62–80

9 DENSITY ESTIMATION AND THE VISUALIZATION OF POINT PATTERNS

A. GATRELL

... INTRODUCTION

The widespread availability of various commercial GIS, coupled with the wealth of point data in digital form, has led to a demand for new methods to represent and analyse such data. Examples of point objects, each referenced by a pair of locational coordinates, are not hard to find. Particular attention is given in this chapter to social data, especially that of an address-based nature, and the examples chosen refer to unit postcodes in Britain, but the ideas generalise to other contexts. Although in general postcodes represent clusters of households rather than individual properties, they have attached to them single grid references, the set of which may be treated as a point pattern.

If we have a digital database containing point data it is a trivial exercise to map these data as a 'dot map'. These can give a reasonable impression of variations in the density of such objects across the study region. Yet, given that density is a spatially continuous variable, it is useful to explore methods that will transform a dot map of points into a continuous density surface. The method of 'density estimation' described here provides a potentially useful visualization of any point data. In GIS, the idea that it is possible to transform one class of object into another is not new (Goodchild, 1985; Gatrell, 1991).

In population mapping, the idea of deriving a continuous surface from a set of discrete points has been proposed before (Kolberg, 1970). Moreover, despite lengthy debates in the literature, it is now generally accepted that while households are discrete objects population density is a spatially continuous function, with a value at any location within an appropriate study region (Nordbeck and Rystedt, 1970). More recently, Martin (1988; 1989) and Bracken (see Chapter 10) have developed techniques for population mapping that avoid dependence on essentially arbitrary areal units.

The present work has its origins in research into disease clustering (Diggle, 1990; Diggle et al., 1990) in which the aim was to assess whether the incidence of a rare cancer was associated with proximity to a possible point source of pollution. The model used required estimation of background population distribution from point data (data on other, more common, cancers), and also data on the distribution of unit postcodes. The approach used 'kernel' methods which, although relatively new in spatial analysis, are widely used in statistics (Silverman, 1986). We outline these methods in a later section.

...POPULATION: ESTIMATION AND VISUALIZATION

Much traditional population mapping using point objects has relied on representing population totals by symbols such as circles or spheres (Robinson et al., 1985, 291–6), though it is now possible to automate production of such maps (Rase, 1987). Alternatively, we can use dot maps (Robinson et al., 1985, 300–306), whereby each dot represents a particular amount, or unit value, of the phenomenon being mapped. The former method assumes that the data have already been aggregated to some set of areal units, but is useful for representing population totals rather than density. The latter requires careful selection of the unit value and some arbitrariness in the choice of locations.

In an important paper Nordbeck and Rystedt (1970) produced a lengthy justification of the use of isarithmic maps in population mapping (see Ottoson and Rystedt, 1980, for a brief summary). Their work anticipates by twenty years contemporary work on GIS and address-based data, arguing that 'the location of every individual's dwelling should be known' (Nordbeck and Rystedt, 1970, p. 107). Although in their own empirical work they were forced to use areally aggregated data, they referred to 'advanced plans concerning the assignment of coordinates to the Swedish road and street systems' (Nordbeck and Rystedt, 1970, p. 108). In terms of the present chapter, their article is of primary importance both because it formalises the notion of a (population) density function and because it anticipates the methods described below. 'The population density function f(x,y) is simply defined as the number of individuals living in a reference area, located around points (x,y)' (Nordbeck and Rystedt, 1970, p. 112). Moreover, they argue that these reference areas (which might typically be circular in shape) must be overlapping. As they further observe, increasing the size of the reference area serves to smooth the estimate of population density.

Examples of the visual representation of population density using surface methods are still not very common. Haggett et al. (1977, p. 419) discuss the classic work by Matui in Japan on the distribution of houses in the Tonami Plain and map both the raw data and isarithms of dwelling density. Much more common is to see population-based choropleth maps derived from census data. These data are available at a variety of spatial scales (electoral wards or Enumeration Districts in Britain, for example). Such choropleth maps have been criticised by Martin (1988; 1989) and alternative methods proposed. One method (Chapter 10) transforms data from irregularly shaped areal units to a raster grid. Another transforms the areal units into a cartogram representation (Chapter 11), with the areas of units (such as circles) proportional to population size or any other variable of interest.

Martin's technique for population mapping is based on centroids for small areas (Enumeration Districts in Britain). The method takes each centroid in turn and distributes the population or other count data attached to that point around the local area. A model is constructed that allocates to any cell an estimated population; this estimate is a weighted sum of the population attached to those centroids lying within a neighbourhood of the cell. The weights are based on some function of distance from the centroid to the cell. The maximum extent of the neighbourhood depends upon the density of centroids, such that a centroid lying in a dense cluster may contribute population only to the cell within which it lies, while the population of an isolated centroid is likely to be spread among several cells.

... DENSITY ESTIMATION

The approach adopted in this chapter bears some similarity to that of Martin and Bracken. The principles are easier to grasp in one, rather than two, dimensions. We, too, use discrete point locations as the basic source of data. The main difference from related work reported in this volume is that while Bracken uses information on the population sizes of areal units (albeit assigned to point locations), our work uses only locational information. In this sense, the work outlined subsequently produces maps not of population density but of residential density (or 'risk' in the case of epidemiological or crime data, for example).

The notion of a probability density function (pdf) is central in inferential statistics. The concept of a pdf is quite easily grasped through a study of the normal distribution, a continuous distribution with parameters representing the mean and variance. We sometimes need to fit such a pdf to observed data; this amounts to estimating the mean and variance, which we may then insert into the formula for the normal density function in order to represent it graphically (Figure 9.1). Such a pdf is parametric; there is, however, a large and growing literature in statistics on nonparametric density estimation, in which rigid assumptions about the nature of the underlying distribution are not required.

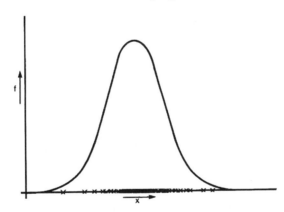

Figure 9.1 A normal probability density function

As Silverman (1986) observes, the most widely used density estimator for univariate data is the histogram, in which we count the relative frequency of objects in each of a series of classes or 'bins'. However, very different representations of histograms can be obtained, depending upon the location of the origin and the class intervals or 'bin width'. Further, while two-dimensional histograms (sometimes called 'Manhattan plots') can be produced using graphics packages, these are subject to the same problems, with the further difficulty that the orientation of the bins is often arbitrary.

Consequently, a class of density estimators called 'kernel methods' has come to be adopted. In general, and following Silverman's notation, we can write a kernel estimator as:

$$f(x) = \frac{1}{nh} \sum^{i} K\left(\frac{X-X_i}{h}\right) \qquad ...(1)$$

where x is the variable of interest and X_i is an observed measurement of x for case i. The number of observations is given by n and the parameter h is a value which controls the amount of smoothing of the density. The choice of kernel function K(.) is to some extent arbitrary; it might, for instance, be a normal (Gaussian) function. The consensus in the literature (Silverman, 1986) is that the choice of kernel function is much less critical than the choice of the smoothing parameter ('bandwidth' or 'window width').

As Figure 9.2(a) makes clear, the estimator is the sum of a series of individual kernels centred over each data point in turn. Given any location x, therefore, we can estimate density using equation (1). We can also see (Figures 9.2(b) and 9.2(c) how results are a function of h. As this is increased more detail in the density estimate is obliterated, while the smaller the h the less the smoothing and the more 'spiky' the density estimate becomes; if the data are undersmoothed then spurious structures are introduced into the density estimation. Notice that in all three illustrations the data points remain unchanged. Clearly, some rules are needed concerning the selection of the smoothing parameter. Diggle (1985) has explored a method for estimating it when the data are generated by a particular point process in one dimension and this too can be carried over into two dimensions.

To explain Diggle's method for selection of the smoothing parameter requires considerable knowledge of the theory of stochastic point processes and we simply make use of the results of this theory below. Briefly, however, we can assume that there is an underlying process that generates points and that there exists an unknown density (called 'intensity' in the point process literature) $\Lambda(x)$, at any location x. We can estimate such an intensity, which we call $\lambda(x)$, using equation (1), assuming we have a value for h. The mean square error (MSE), itself a function of h, relates $\Lambda(x)$ to $\lambda(x)$ and, when estimated, gives a set of values for any h. We can then select that value of h which minimises MSE. This method was used in earlier epidemiological work (Diggle et al., 1990) that sought to model raised incidence of laryngeal cancer around the site of a former industrial waste incinerator.

Brunsdon (1991), who also outlines this method of density estimation, follows Martin and Bracken in suggesting that the bandwidth parameter should be spatially variable; in other words, if there are substantial variations in density across the study region h should be modified to reflect this, with a larger value in areas of low density and smaller value where observations are clustered. This is known as 'adaptive' kernel estimation.

The method of density estimation, and Diggle's approach to estimating the smoothing parameter, are both extendable to two-dimensional point processes. We now give some illustrations.

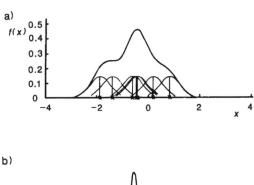

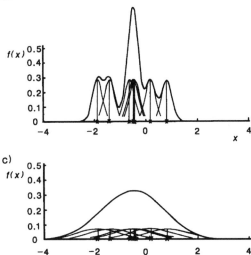

Figure 9.2 Density estimation: effect of varying smoothing parameters

...APPLICATIONS OF DENSITY ESTIMATION IN TWO DIMENSIONS

Brunsdon (1991) gives an example of density estimation using grid-referenced crime data (household burglaries). As he notes, the resulting three-dimensional plots show 'hot-spots' of crime incidence and may be overlain with other datasets to provide contextual information. Bithell (1990) uses the method in a study of childhood leukaemias in Cumbria. He obtains a surface for both the leukaemia data ('cases') and for a set of randomly selected addresses ('controls') and forms an estimate of 'relative risk' that assesses whether the incidence of the disease is raised locally. The surface shows clearly the well-known peak around Sellafield, though the degree of clarity does vary with how the bandwidth parameter is selected.

Here, we outline a further example relating to address-based data for the London Borough of Camden. Two sets of points are considered. One is taken from the Central Postcode Directory (CPD), the other from data supplied for research purposes by Pinpoint Analysis Ltd. The CPD is a file comprising about 1.5 million records, listing each unit postcode in the country, together with a 100 metre resolution OS grid reference. Such grid references have in general been assigned clerically by the Post Office and allocate the first property in the postman's walk to the south-west corner of the 100 metre grid square in which it lies. Other work in the North West Regional Research Laboratory (Gatrell, 1989; Gatrell et al., 1991) has assessed the accuracy of this spatial referencing.

We extracted from the CPD all unit postcodes that lie within Camden District. Only postcodes for 'small users' of mail have been extracted, that is, ignoring premises and large organisations which might handle substantial volumes of mail. There are 4490 such unit postcodes in Camden. Given the 100 metre resolution of the data we obtain the 'raster-like' plot of the grid references shown as Figure 9.3. It should be borne in mind that several postcodes might be assigned to the same grid reference; a good reason why such a point map does not itself represent density very faithfully.

Figure 9.3 Distribution of unit postcodes from the CPD in Camden

Pinpoint Analysis Ltd have been engaged in digitising all properties and roads in the country. All properties are assigned a grid reference accurate to 1 metre, forming a file known as the Pinpoint Address Code (PAC). As part of their association with the Regional Research Laboratory initiative (see Masser, 1990), they made available to all RRLs the file of digitised addresses for Camden. This comprises 31 134 records. All properties with the same unit postcode have been collected together in the present work and an average grid reference obtained. We refer to this below as the PAC centroid and these are mapped in

Figure 9.4. We thus have two sets of 4490 matched point locations. A detailed comparison of these is reported elsewhere (Gatrell et al., 1991). We simply report here on the use of both sets of point events to visualize surfaces of post-code 'intensity'.

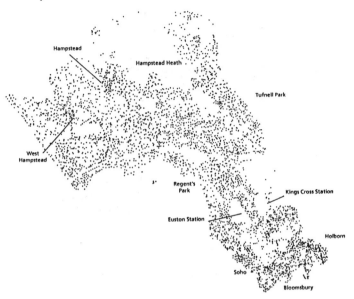

Figure 9.4 Distribution of PAC centroids in Camden

For convenience, the grid references have been converted to lie within a unit square. For the CPD data the smoothing parameter that minimises mean squared error is estimated as 0.02, corresponding to a distance of 200m on the ground. For the PAC data the value is 0.01 corresponding to a distance of 100m. The resulting maps of intensity, shown as Figures 9.5 and 9.6, are, of course, broadly similar, but differ in detail. High densities in Hampstead and West Hampstead in the north west are picked out in both maps, as they are in Holborn, Bloomsbury and towards Soho in the south. The surface derived from the PAC data, however, offers a more finely grained characterisation of density variation. In particular, more local peaks and depressions in the density surface are identified in the PAC map than the CPD. Of course, in general, only data from the CPD are in the public domain; individual PAC data are only available on a commercial basis. The results shown here, however, suggest that useful information on density variation is available from the CPD. Both provide a useful visualization alternative to the conventional dot map.

... SOFTWARE ENVIRONMENTS

How do we implement such density estimation? Currently, this is done at Lancaster in three ways. First, FORTRAN code is available to perform, separately, the estimation of the smoothing parameter and the estimation of intensity at a

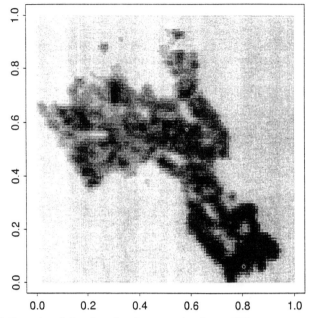

Figure 9.5 Contoured density of unit postcodes from CPD

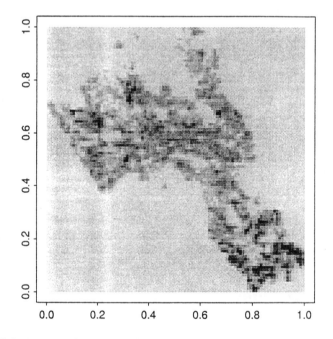

Figure 9.6 Contoured intensity of PAC centroids

set of discrete locations. The latter may then be contoured within any suitable package (e.g. UNIMAP). This is how the Camden illustrations (Figures 9.5 and 9.6) were produced. Second, and perhaps more usefully, we can use an AML program within ARC/INFO; a separate call to external FORTRAN routines performs the estimation of the smoothing parameter and estimation of intensity and the results are imported into ARC/INFO for further processing and display. This is part of a wider research programme, in collaboration with the North East RRL, to infuse ARC/INFO with more spatial analytical capability (see Rowlingson et al., 1991, for some other examples). Third, we can use the statistical programming language S-Plus. Colleagues at Lancaster (Rowlingson and Diggle, 1991) have shown how to perform point process modelling within S-Plus, by adding new functions that are compiled FORTRAN subroutines. Density estimation using kernel methods is one such function, the results of which can be displayed within S-Plus as raster output or as a three-dimensional model. A disk for performing this and other spatial point process tasks is available for those working in a UNIX environment; see Rowlingson and Diggle (1991) for further details.

... CONCLUSIONS

The above results suggest that it is feasible and worthwhile to use the locations derived from the Central Postcode Directory to create surfaces that, in a general sense, serve as maps of the spatial variation in density. The methodology is sufficiently general as to be applicable in a range of situations. It is of considerable value in epidemiological or crime mapping and analysis, where we frequently wish to estimate surfaces showing geographical variation in the incidence of diseases or crimes, and when we wish to go beyond this to test hypotheses about environmental associations.

... ACKNOWLEDGEMENTS

This work follows collaborative research with Professor Peter Diggle (Mathematics, Lancaster) and would have been impossible without access to his algorithms, expertise and help, all freely given. Much of the research relies on access to data from Pinpoint Analysis Ltd, made available by the company to the ESRC-funded Regional Research Laboratories. Colleagues within the North West Regional Research Laboratory, particularly Robin Flowerdew, Barry Rowlingson and Isobel Naumann, provided help and intellectual stimulation.

... REFERENCES

Bithell J F (1990) An application of density estimation to geographical epidemiology. *Statistics in Medicine* 9: 691–701

Bracken I, Martin D (1989) The generation of spatial population distributions from census centroid data. *Environment and Planning A* 21: 537–43

Bracken I, Webster C (1990) *Information Technology in Geography and Planning*, Routledge, London

Brunsdon C (1991) Estimating probability surfaces in GIS: an adaptive technique. In: Harts J, Ottens H F L, Scholten H J (eds) *Proceedings, Second European Conference on Geographical Information Systems* EGIS Foundation, Utrecht, Netherlands, pp. 155–64

Diggle P J (1985) A kernel method for smoothing point process data. *Applied Statistics* 34: 138–47

Diggle P J (1990) A point process modelling approach to raised incidence of a rare phenomenon in the vicinity of a pre-specified point. *Journal of the Royal Statistical Society, Series A* 153: 349–62

Diggle P J, Gatrell A C, Lovett A A (1990) Modelling the prevalence of cancer of the larynx in part of Lancashire: a new methodology for spatial epidemiology. In: Thomas R W (ed.) *Spatial Epidemiology*, Pion, London, pp. 35–47

Gatrell A C (1989) On the spatial representation and accuracy of address-based data in the United Kingdom. *International Journal of Geographical Information Systems* 3: 335–348

Gatrell A C (1991) Concepts of space and geographical data. In: Maguire D J, Goodchild M F, Rhind D W (eds) *Geographical Information Systems: Principles and Applications*, Longman, Harlow, pp. 119–34

Gatrell A C, Dunn C E, Boyle P J (1991) The relative utility of the Central Postcode Directory and Pinpoint Address Code in applications of geographical information systems. *Environment and Planning A* 23: 1447–1458

Goodchild M F (1985) Geographical information systems in undergraduate geography: a contemporary dilemma. *The Operational Geographer* 8: 34–38

Haggett P, Cliff A D, Frey A E (1977) *Locational Methods in Human Geography*, Edward Arnold, London

Kolberg D W (1970) Population aggregations as a continuous surface. *Cartographic Journal* 7: 95–100

Martin D (1988) An approach to surface generation from centroid-type data. *Technical Reports in Geo-Information Systems, Computing and Cartography* 5, Wales and South West Regional Research Laboratory, Cardiff UK

Martin D (1989) Mapping population data from zone centroid locations. *Transactions, Institute of British Geographers* 14(1): 90–97

Martin D, Bracken I (1991) Techniques for modelling population-related raster databases. *Environment and Planning A* 23: 1069–1075

Masser I (1990) The Regional Research Laboratory initiative: an update. In: *The Association for Geographic Information Yearbook*, Taylor and Francis, London, pp. 259–63

Nordbeck S, Rystedt B (1970) Isarithmic maps and the continuity of reference interval functions. *Geografiska Annaler* 52B: 92–123

Ottoson L, Rystedt B (1980) Computer-assisted cartography: research and applications in Sweden. In: Taylor D R F (ed.) *The Computer in Contemporary Cartography*, Wiley, Chichester, pp. 93–122

Rase W D (1987) The evolution of a graduated symbol software package in a changing graphics environment. *International Journal of Geographical Information Systems* 1: 51–66

Robinson A H, Sale R, Morrison J, Muehrcke P (1985) *Elements of Cartography*, 5th edn, Wiley, Chichester

Rowlingson B S, Diggle P J (1991) SPLANCS: spatial point pattern analysis code in S-Plus. *North West Regional Research Laboratory, Research Report 22*, Lancaster University, Lancaster UK

Rowlingson B S, Flowerdew R, Gatrell A C (1991) Statistical spatial analysis in a Geographical Information Systems framework. *North West Regional Research Laboratory, Research Report 23*, Lancaster University, Lancaster UK

Silverman B W (1986) *Density Estimation for Statistics and Data Analysis*, Chapman and Hall, London

10 Towards Improved Visualization of Socio-Economic Data

I. Bracken

... Introduction

The development of computer graphics, coupled with that of database management systems, has given scientists in many fields powerful new techniques for the graphic representation and visualization of their material. Moreover, the speed of data processing now means that changes can be readily portrayed and the results of experiments on data, for example by the manipulation of a model, can be visualized in real time. As widely noted these developments have given analysts persuasive new tools (Teicholz and Berry, 1983). However, they also highlight the need for greater attention to the validity and meaning of the visual message, for computer graphics can be uniquely persuasive as a communicating medium (Robbins and Thake, 1988).

For geographers, the visual representation of spatial information on maps, diagrams and charts has long been an integral part of their science (Robinson and Petchenik, 1975). Indeed, a correct representation of the real world is fundamental to all forms of geographic study, yet the process of representing geographic reality in a scientific way is complex, involving a careful balancing of a number of conceptual and practical issues (Forbes, 1984; Langford et al., 1990b; Trotter, 1991). Although representation techniques and theory are critically discussed in the literature from time to time (for example by Shirayev, 1987; Peuquet, 1988), there has been inadequate recognition of the importance of such issues in substantive studies (Openshaw, 1984; Chrisman, 1987). Nowadays, the increasing availability of proprietary software for geographic analysis and display has a tendency to exacerbate this problem because issues of data quality and representation are largely hidden from users, whose aim is often quite pragmatic. All too frequently, it is assumed that those who developed the techniques have also faced and solved underlying theoretical issues. Although there is evidence that many users are increasingly aware of this issue, and certainly the theme is beginning to be more widely addressed in applied literature, it remains the case that much of the visual representation of geographic information is conventionally and uncritically employed.

Choropleth maps are widely used for the graphic portrayal of social data from census and surveys. Frequently, the data acquisition unit (e.g. a census zone) is regarded as satisfactory for the display and visualization of such data but in most cases these zonal arrangements are imposed arbitrarily onto spatial distributions for administrative, or simply 'data collection', convenience. Once adopted, such structures tend to persist and indeed they often dominate subsequent analyses. There are considerable representational dangers in the uncritical use of such data, the more so as many proprietary GIS use comparatively unsophisticated mapping routines which invite the linking of attribute data to digitised map zones. The key question, often far from adequately addressed, is about the appropriateness of the zonal representation in terms of the spatial distribution of the information being portrayed. This is particularly an issue in the socioeconomic domain where zone-based data enumeration is commonplace.

Data manipulation and transformation techniques within GIS provide possibilities for the re-representation of data in different forms. These principles have been well understood for many years and improved techniques appear regularly (Samet, 1984a, 1984b; Piwowar et al., 1990). A zonal representation may be converted to an equivalent grid form by rasterization and stored efficiently in a quadtree structure, or a gridded database may be converted to a vector-based structure by vectorization or digitization. These transformations may make the data more amenable to display and make for more efficient storage of the information. They do not, however, change fundamentally the form of representation. Nevertheless, such transformations do facilitate an emerging theme of data integration within GIS (Devereux, 1986; Piwowar et al., 1990). Simple techniques for integration include overlay and the generation of 'product layers' in rasterised information structures. Overlaying vector data from a spatially coincidental zonal system is clearly possible, though in all cases there will be some loss of information and errors due to the inherent spatial incompatibility of the zonal structures and these will increase as the zones become more irregular. Nevertheless there is considerable technical work to handle this problem and evaluate the attendant errors of the process (Flowerdew and Green, 1989). Such approaches are based on principles of areal interpolation and may involve the use of supplementary information.

In contrast, there are techniques which explicitly require a data model for the transformation of the data. It is widely acknowledged that, in both data visualization and data integration, the use of a georeferenced grid has many useful properties (Tobler, 1979; Browne and Millington, 1983; Kennedy and Tobler, 1983). In this chapter the transformation of socioeconomic data into such a grid form via a data model is explored. It should be noted that this process is *not* one of the rasterisation of an existing zonal structure. Indeed, the approach requires a quite different assumption about the basic spatial meaning of a 'unit of information'. Some geographic data are already collected in grid form, the most prominent being remotely sensed information. Even here the data are normally transformed by geometric re-sampling onto a standard grid, or to align the data with other information systems such as the UK Ordnance Survey National Grid. Well-established transformations are available in raster-based GIS to effect these transformations with minimal, known information loss.

Where data are not collected in a grid form other types of transformation are called for and a possible model for a *set of data* that can be validly related to georeferenced points, such as a set of census zone centroids or postcode locations, is now identified in terms of its visualization properties. In effect the approach is to generate a surface of social data values by making assumptions about their distribution over space.

...A SURFACE MODEL FOR SOCIOECONOMIC DATA

It can be argued that the representation of population and population-related phenomena, which are punctiform in nature, can best be done using a structure which is independent of the peculiar and unique characteristics of the actual spatial enumeration zones. One way in which this can be achieved is to transform zone-based census data and point-based address data into a data structure that represents population distribution more plausibly as something approaching a continuous surface. To operationalise this approach requires a change in a basic assumption of data representation, namely that the data value (the count) is seen not as a property of an area or a point, but rather as a property of a location at which the population (and related phenomena) are distributed. It is then further assumed that useful information about this distribution can be derived from a local, spatial analysis of a *set* of the basic area or point data values. Using this assumption, a model can be created to generate a spatial distribution of population (or households) as a fine, variable resolution geographical grid with assessable error properties (Bracken and Martin, 1989).

In the UK, machine readable georeferences are available for both postcode units and census zones, all related to the Ordnance Survey National Grid. For the former, a single file contains all residential unit postcodes and their grid locations to a 100 metre resolution (Gatrell et al., 1991; see also Chapter 9). In practice, this means that a small number of codes share a common grid reference. In the case of the census, every enumeration district has assigned to it a georeferenced centroid which is population-weighted by being manually located at the approximate centre of the populated part of the zone. Again, this information is recorded to the nearest 100m. The spatial pattern of census centroids or postcode locations will be quite different for an isolated settlement, a village, a linear settlement, a suburb or the central part of a city. One such location cannot say anything about the local distribution of population, but, as shown in Chapter 9, a kernel-based analysis can be used to generate a plausible distribution according to the local density of a given set of point data locations. This uses a derivative of moving kernel density estimation (Bowman, 1973; Silverman, 1986; Martin, 1989) and can be shown to be capable of both reconstructing the settlement geography from the georeferenced population 'points' and of preserving the volume (ie. the count) under the density surface.

In the analysis, a window is positioned over each data point in turn, and the size of this window is varied according to the local density of surrounding points. In effect, this provides an estimate of the size of the areal unit assumed to be represented by the current location, whether census centroid or postcode. The count associated with this location is then distributed into the cells falling within the window, according to weightings derived from a distance-decay function, as used by Cressman (1959). This function, which here uses a finite window size, is formulated to increase the weight given to cells nearest the data point in relation to local point density (Martin and Bracken, 1991). The allocation involves only integer (ie. whole person) values which means that the distri-

bution process has finite, and in practice quite local, bounds thereby reconstructing a good estimate of populated and unpopulated areas.

In summary, the assignment to each cell in the grid is obtained as:

$$P_i = \sum_{j=1}^{c} P_j . W_{ij}$$

where P_i is the estimated population of cell i, P_j is the empirical population
recorded at point j, c is the total number of data points, and W_{ij} is the unique
weighting of cell i with respect to point j. Cell i will not receive population
from every point location but only from any points in whose kernel it falls (W_{ij}
will be zero for all other cell-point combinations). At each location, the kernel
is initially defined as a circle of radius r, and is locally adjusted according to the
density of other points falling within distance r of that location j. Weights are
then assigned to every cell whose centre falls within the adjusted kernel,
according to the distance decay function:

$$W_{ij} = f \left(\frac{d_{ij}}{r_j} \right)$$

where f is the distance decay function; d_{ij} is the distance from cell i to point j,
and r_j is the radius of the adjusted kernel defined as:

$$r_j = \frac{\sum_{l=1}^{k} d_{jl}(j=1)}{k}$$

where $1 = 1, 2, ..., k$ are the other points in the initial kernel window of radius
w. If there are no other points with $d_{ij} < w$ (ie. k=0), then the window size w_j
remains set equal to w. This sets the dispersion of population from point j to a
maximum in the case where there are no other points within the entire window. In effect, the value r determines the maximum extent to which population will be distributed from a given location. Where k > 0, a greater clustering
of points around point j (smaller radius r_j) will result in a smaller window and
hence greater weights being assigned to cells falling close to j. In the most
localised case, where r_j is less than half the size of a cell, it is probable that the
cell containing the location will receive a weight of 1, and the entire population at that location will be assigned to the cell. It should also be noted that the
weight given to a cell centre spatially coincident with the location j also cannot
be greater than 1, as the location must then contribute population to the output grid exactly equivalent to its attribute value. Once weights have been
assigned to every cell in the kernel, they are re-scaled to sum to unity thereby
preserving the total population count P_j to be distributed from location j. As
some cells fall within the kernels based on several different points, they receive
a share of the population from each. In contrast, other cells, beyond the region
of influence of any points, remain unvisited and receive no population and
hence define the unpopulated areas in the complete raster surface.

In early implementations of this model (Martin, 1989) a simple distance decay function was used in which the weighting was directly and inversely related distance from each input location. Latterly, a more flexible alternative has been developed to define the weightings as:

$$w_{ij} = \left[\frac{w_j^2 - w_{ij}^2}{w_j^2 + d_{ij}^2} \right]^a (d_{ij} < w_j)$$

This function has a value of zero when d_{ij} is greater than or equal to r_j. Increasing the value of a increases the weight given to cells close to the data point thereby achieving greater control over the form of the distance decay.

Using these techniques, a highly plausible raster model has been implemented which is capable of providing a good approximation to a spatially continuous estimate of population density from census or postcode sources. The importance of being able to assess the errors of the process should not be overlooked (Goodchild, 1989). Although the model appears to work well with UK census and postcode data, it is recognized that all processes involving kernel-based estimation will be sensitive to the bandwidth of the kernel for which evaluation techniques have more generally been suggested (Bowman, 1984).

Using this model a series of population-related raster databases have been generated from zone-based 1981 British Census data at fine resolution and for extensive areas. National layers in the database use a 200m cell size giving some 330 000 populated cells out of the 16 million possible cells in the overall national grid. A number of key census variables have been modelled onto the same grid, and spatial indicators can readily be computed at the level of individual cells using the population (or household) surface as the denominator. As the density of postcode locations is greater than the density of census zone centroids, the model will support a much finer resolution grid where data are based on the unit postcodes. It should be noted that the two data sources can be used in conjunction, for example the postcode locations can be used to generate a finer spatial distribution of the census zone population counts. In this case, the census provides the counts and the postcodes provide the distributional information via the model.

...VISUALIZATION OF SOCIAL DATA

Having generated a raster database of census variables, the next step is to visualize these data. There has been long-standing discussion about the representation of population in geographical analysis (Mackay, 1951; Schmid and MacCannell, 1955; Nordbeck and Rystedt, 1970; Langford et al., 1990a) and a recognition that population density, and population-related phenomena such as unemployment, could be viewed as continuously varying (Unwin, 1981). More recently such recognition has been stimulated by interest in the representation of population as a surface by using satellite sources on settlement information (Iisaka and Hegedus, 1982; Lo, 1989; Langford et al., 1990b) and set within the more general ideas on the continuous representation of geographic phenomena (Trotter, 1991). Whatever form the representation takes, it has been shown by Peuquet (1988) to depend upon the validity of the conceptual links between the three key components of a representation process briefly noted earlier. These components are, first, the concept of the phenomenon (ie. the definition of 'population'); second, how it can be observed and measured (ie. counts of people at particular locations); and third, the data model (ie. the form in which the concept is represented by the data, both in locational and substantive terms). In transforming data from one form of representation to another, all the components of this process and their linking need to be fully recognised.

Where population information is not directly captured in raster form, and this will normally be the case with social survey, the data are usually aggregated and made available at the level of the irregular reporting or enumeration units. Although 'simple' dot mapping these data is possible, it is subject to limitations relating to scale. Maps of large areas can simply use the centroid of each zone as a locating point for each displayed datum, but, as the scale of visualization becomes more local, the quality of representation of the underlying phenomena and its spatial distribution begins to break down. In particular, the irregular geography of the zonal system itself begins to mask the real, underlying distribution of population. Interpretations of resulting displays are therefore highly dependent upon scale of visualization in relation to the zonal structure of the data. Clearly, the transformation of data into a raster database by means of a surface model is one way to address this difficulty. The resulting database is not only able to support visualization at a very wide range of scales, but also has valuable properties for spatial analysis.

Where truly continuous phenomena (such as the land surface), or punctiform phenomena (such as population) are to be represented in locationally discrete data structures, some generalisation is always necessary. Raster data models have proved to be an appropriate way of approximating such continuous variation (Peucker et al., 1978) and are well established in similar fields such as altitude matrices in digital elevation models (Yoeli, 1986; Theobald, 1989; see also Chapter 18). An idealised spatial population database would take the form of a population register with individual georeferencing, or perhaps a census with household georeferencing. Practical and societal limitations (e.g. confidentiality) make such data rare but it is worth noting that a very fine resolution raster database could result. The calculation of population density would be possible around very many points within a local area, and the resulting surface would have properties close to an idealised continuous representation.

For visualization, the issue of scale becomes important particularly in regard to a concept such as population density which can be portrayed directly in raster form where the spatial unit of enumeration and reporting permits. For example, the British Population Census of 1971 allocated the population to 1km grid squares, allowing the display of census results as a raster-based atlas (Census Research Unit, 1980), a development in some ways ahead of its time in the use of socioeconomic information. More recently, the results of the 1990 Swedish Population Census have been portrayed in a similar manner with grids at a variety of resolutions covering both extensive and local areas (Oberg and Springfeldt, 1991). For both of these applications, the data processing required was simply one of spatial aggregation from the georeferenced person or household to the relevant grid.

An important feature of a generated surface at the resolution described is that the underlying settlement pattern is clearly revealed by mapping the cells. From a single surface layer a topic (say the proportion of the elderly population) can be readily mapped from the national scale down to the local scale with unique flexibility. In the work described here, these outputs are achieved at a high resolution by the use of the University of London's computer-driven film plotter attached to its Convex processor. Specially written software provides complete flexibility in terms of geographic coverage and colour. In addition, digital source files covering the coastline, county, district and ward boundaries are available to be added as vector overlays to provide help in location identification.

Plates 3 to 6 illustrate a range of scales and the effective reconstruction of settlement patterns. The grid base for all these plates is the standard 200m cell used in the model and the data all relate to the 1981 Census. Each variable was processed separately and comprises a single layer in the database. In all cases, the population and household surfaces were modelled independently for the basic census counts providing in effect denominator surfaces and the counts for each variable were then modelled in the same way to provide the numerator. The product surfaces visualized here are the results of cell-by-cell computations, for example the proportion of elderly, and so on. Plate 3 shows the percentage of population aged 60 and over for London and the South East. Plate 4 shows the percentage of non-car owning households in the Birmingham/ Coventry area; Plate 5 the percentage of population seeking work at the time of the census (ie. 'unemployed' over South Wales); and Plate 6 shows at a larger scale the percentage of population who lived at a different address one year prior to the census in the City of Cardiff.

... CONCLUSIONS

In this chapter the idea of transforming data via a data model into a structure that performs better as a visualization tool, introduced in Chapter 9, has been further explored. Once again, this transformation has implied facing head-on a number of difficult issues to do with the most appropriate representation of socioeconomic information and the interpretation of graphic portrayal. It has been suggested that the georeferenced grid has considerable potential as a general data structure for geographical data of many types, not least because it can facilitate the integration of data and the added value of information which this implies. This seems particularly so in regard to the use of visualization techniques for the understanding of social indicators (Bracken 1989, 1992). The technique described here is appropriate in the UK with its unique system of census population-weighted centroids, though the application of the technique to georeferenced postcodes has a more general utility. The approach is suggested as one way in which the visualization of social information can be enhanced by transformation with known model and error properties.

... REFERENCES

Bowman A W (1973) A comparative study of some kernel-based nonparametric density estimators. *Journal of Statistical Computation and Simulation* 21: 313–327

Bowman A W (1984) An alternative method of cross-validation for the smoothing of density estimates. *Biometrika* 71: 353–360

Bracken I (1989) The generation of socioeconomic surfaces for public policy making. *Environment and Planning B* 16: 307–326

Bracken I (1991) A surface model of population for public resource allocation. *Mapping Awareness* 5(6): 35–39

Bracken I (1992) A surface model approach to the representation of population-related social indicators. Paper presented to workshop I-14, San Diego. National Center for Geographic Information and Analysis, University of Buffalo, New York

Bracken I, Martin D (1989) The generation of spatial population distributions from census centroid data. *Environment and Planning A* 21: 537–543

Browne T J, Millington A C (1983) An evaluation of the use of grid squares in computerised choropleth maps. *The Cartographic Journal* 20: 71–75

Census Research Unit (1980) *People in Britain – A Census Atlas*, HMSO, London

Chrisman N R (1987) Fundamental principles of geographic information systems. In: *Proceedings Auto Carto 8*, American Society for Photogrammetry and Remote Sensing, Falls Church Virginia, pp. 32–41

Cressman G P (1959) An operational objective analysis system. *Monthly Weather Review* 87(10): 367–374

Devereux B J (1986) The integration of cartographic data stored in raster and vector formats. In: *Proceedings, Auto Carto London*, Royal Institution of Chartered Surveyors, London, pp. 257–266

Flowerdew R, Green M (1989) Statistical methods for inference between incompatible zonal systems. In: Goodchild M F, Gopal S (eds) *Accuracy of Spatial Databases*, Taylor and Francis, London, pp. 239–248

Forbes J (1984) Problems of cartographic representation of patterns of population change. *The Cartographic Journal* 21(2): 93–102

Gatrell A C, Dunn C E, Boyle P J (1991) The relative utility of the Central Postcode Directory and Pinpoint Address Code in applications of geographical information systems. *Environment and Planning A* 23: 1447–1458

Goodchild M F (1989) Modeling error in objects and fields. In: Goodchild M F, Gopal S (eds) *Accuracy of Spatial Databases*, Taylor and Francis, London, pp. 107–114

Iisaka J, Hegedus E (1982) Population estimation from Landsat imagery. *Remote Sensing of Environment* 12: 259–272

Kennedy S, Tobler W R (1983) Geographic interpolation. *Geographical Analysis* 15(2): 151–156

Langford M, Unwin D J, Maguire D J (1990a) Generating improved population density maps in an integrated GIS. *Proceedings of the First European Conference on Geographical Information Systems*, EGIS Foundation, Utrecht

Langford M, Unwin D J, Maguire D J (1990b) Mapping the density of population: continuous surface representations as an alternative to choroplethic and dasymetric maps. *Research Report* 8, Midlands Regional Research Laboratory, Leicester UK

Lo P (1989) A raster approach to population estimation using high-altitude aerial and space photographs. *Remote Sensing of Environment* 27: 59–71

Mackay J R (1951) Some problems and techniques in isopleth mapping. *Economic Geography* 27(1): 1–9

Martin D (1989) Mapping population data from zone centroid locations. *Transactions, Institute of British Geographers* 14(1): 90–97

Martin D, Bracken I (1991) Techniques for modelling population-related raster databases. *Environment and Planning A* 23: 1069–1075

Nordbeck S, Rystedt B (1970) Isarithmic maps and the continuity of reference interval functions. *Geografiska Annaler* 52B: 92–123

Oberg S, Springfeldt P (eds) (1991) *The Population Atlas of Sweden*, SNA, Stockholm

Openshaw S (1984) *The Modifiable Areal Unit Problem,* Concepts and Techniques in Modern Geography 38, Geo Books, Norwich UK

Peucker T K, Fowler R J, Little J J, Mark D M (1978) The triangulated irregular network. *Proceedings of the DTM Symposium*, American Congress on Surveying and Mapping, St Louis, Missouri, pp. 24–31

Peuquet D J (1988) Representations of geographic space: toward a conceptual synthesis. *Annals, Association of American Geographers* 78: 375–394

Piwowar J M, LeDrew E F, Dudycha D J (1990) Integration of spatial data in vector and raster formats in geographical information systems. *International Journal of Geographical Information Systems* 4(4): 429–444

Robbins R G, Thake J E (1988) Coming to terms with the horrors of automation. *The Cartographic Journal* 25: 139–142

Robinson A H, Petchenik B B (1975) The map as a communication system. *The Cartographic Journal* 12: 7–14

Samet H (1984a) Algorithms for the conversion of quadtrees to raster. *Computer Graphics and Image Processing* 26: 1–16

Samet H (1984b) The quadtree and related hierarchical data structures. *ACM Computing Survey* 16: 187–260

Schmid C F, MacCannell E H (1955) Basic problems, techniques and theory in isopleth mapping. *Journal of the American Statistical Association* 50: 220–239

Shirayev E E (1987) *Computers and the Representation of Geographical Data,* Wiley, Chichester

Silverman B W (1986) *Density Estimation for Statistics and Data Analysis,* Chapman and Hall, London

Teicholz E, Berry B J L (eds) (1983) *Computer Graphics and Environmental Planning,* Prentice-Hall, Englewood Cliffs, NJ

Theobald D (1989) Accuracy and bias in surface representation. In: Goodchild M F, Gopal S (eds) *Accuracy of Spatial Databases*, pp. 99–106

Tobler W R (1979) Smooth pycnophylactic interpolation for geographic regions. *Journal of the American Statistical Association* 74: 519–530

Trotter C M (1991) Remotely-sensed data as an information source for geographical information systems in natural resource management: a review. *International Journal of Geographical Information Systems* 5(2): 225–239

Unwin D (1981) *Introductory Spatial Analysis*, Methuen, London

Yoeli P (1986) Computer executed production of a regular grid of height points from digital contours. *The American Cartographer* 13(3): 219–229

11 CARTOGRAMS FOR VISUALIZING HUMAN GEOGRAPHY

D. DORLING

... VISUALIZATION IN HUMAN GEOGRAPHY

This chapter presents the argument that without the extended use of cartograms, future visualization in human geography will merely repeat a fundamental distortion of much past thematic cartography in (literally) drawing our
attention to the patterns in places where the fewest people live. It is now possible to produce cartograms relatively effortlessly and the author's work on this
is summarised. The important decisions to make today are, first, what form of
cartogram to adopt in a particular situation, and, second, how to best visualize
human geography upon it. First I shall describe what I mean by the terms 'visualization' and 'cartogram'. Next I explain in more detail why we must extend
the use of cartograms, and briefly explain how they can be made. Finally I give
examples of how they can be used in the visualization of human geography.

This book has introduced several definitions of visualization, but each author
will have a slightly different view of so large a subject. To me, visualization
means making visible what was obscure, what could not easily be imagined or
seen. Scientific visualization (ViSC) uses our inherent ability to appreciate a
picture. By transforming large amounts of data into pictures, we can begin to
understand the underlying structure. This approach is based upon the unique
nature of the link between eye and mind. We depend on vision, we think visually, we talk in visual idioms and we dream in pictures. Unfortunately we cannot easily transmit a picture directly from one mind to another, we have to
describe or draw it. Throughout history we have developed (and sometimes
forgotten) ingenious methods of turning numbers and ideas into diagrams and
pictures – turning information into understanding. This process has been most
prolific at those times when the flow of new information was greatest, for
example, at the end of the eighteenth and twentieth centuries.

The first cartograms were created as another way to show just how uneven
was the geographical distribution of population, but later came to be used as a
basis for illustrating the human geography of those populations (Raisz, 1934;
1936). I define cartograms as maps in which a particular exaggeration is deliberately chosen. For example, area cartograms are drawn so that areas on the
paper represent places in proportion to a specific chosen aspect of those places.
Thus an ordinary planimetric map is an equal land area cartogram on which,

projection permitting, areas are drawn in proportion to the amount of land in each place. They are appropriate as a base if you are interested in the spatial distribution of something across the land such as crops. Conversely, in an equal population area cartogram, or population cartogram, areas are proportional to the number of people in each place. A population cartogram is an appropriate basis for seeing how something is distributed spatially across groups of people. Such a cartogram is *not* a distortion of the world, but a representation of some particular aspect of it. Any two-dimensional projection of the surface of the earth must select some aspects to represent and reject others. Almost all ancient maps would today be seen as cartograms, as their exaggerations are seen to be so great, and yet they depicted the reality of their times with uncharted lands compressed and religious capitals in the centre of their worlds (Angel and Hyman 1972). Later on, depicting reality came to mean straightening compass directions to enable trade and conquest over the seas:

> The map is not some inferior but more convenient substitute for a globe. Map projections are not simply choices of lesser evils among distorting possibilities. On the contrary, the map allows the geographer to twist space into the condition he wishes, For purposes of finding lines of constant compass direction, the Mercator projection is far superior to the actual surface of the earth. The earth itself lacks the spatial property of having such lines being straight lines. (Bunge, 1966, p. 238)

Today, depicting reality is still a means to an end, not just to know how to sail around the seas of Europe, but to ask how many people live on each part of the land and in what social conditions. To see the latter more clearly we must begin by using maps in which all groups of people are given equal prominence. The mechanics of this process will now be discussed. Population cartograms giving equal representation to all the people in the picture far outnumber all other kinds produced this century. However, sometimes a selected group from the population would be more relevant (children, households or electors for example). The spatial units used to construct a cartogram are usually politically defined geographical areas but they need not be. The only restriction is that each unit has a known value. However, the units in the cartograms considered here also have known locations and known topological relationships with other units. These are referred to as areal units in the rest of the chapter.

In the 1970s machines were developed to construct population cartograms mechanically, as their manual creation has always been difficult and tedious (Skoda and Robertson 1972). These early methods were not particularly successful in practice and their products were not widely used for other work. By the 1980s work was well underway to create computer algorithms capable of producing useful population cartograms (Tobler 1963, 1976, 1986; Dougenik et al., 1985). Because of the poor visual and cartographic quality of these early attempts most cartograms used today are still generated manually (Eastman et al., 1981), and many from the past have been repeatedly re-used (Hollingsworth, 1964; 1966). This writer has worked on an algorithm to produce cartograms by computer which are acceptable as the basis for future cartographic work. Equal area cartograms of many thousands of areal units can now be generated with ease. The most important current issue is to inform people that usable cartograms exist, and persuade them of their advantages. We need to convince people that using cartograms will broaden their understanding, and that many traditional maps give a false impression of the spatial patterns in human geography. This is not a new debate: '[Cartograms] from many points of view are more realistic than the conventional maps used in geography.' (Tobler, 1961, p. 163) The strongest argument for using a population cartogram rather than a topographic map base is not that the traditional map distorts the pattern across the population, but that most of the pattern is simply not visible! An excellent example is given by Coulson (1977) of a Canadian electoral map in which half the results could not be seen without a magnifying glass. The majority of the population of most countries live in small, densely populated areas which need numerous insets on a national map to give any semblance of justice. This problem becomes more acute as a finer spatial resolution is sought: less than 1 per cent of the (inhabited) 1km grid squares of Britain contained over 30 per cent of the population in 1971 (Craig, 1976). The *People in Britain* 1971 census atlas was an achievement in its time, but the characteristics of the lives of most of the 'people in Britain' were given minimal representation, with those of the sparse rural population dominating the national picture. Data transformations merely served to reduce the arbitrary variations displayed by presenting so many sparsely populated areas so prominently and places with very few people still dominated the map: 'Hence, the maps in "People in Britain" (CRU/OPCS/GRO(S) 1980) are the first reliable maps of unpopulated areas in Britain!' (Rhind, 1983, p. 181)

... THE CHOICES FOR VISUALIZATION

I wanted to draw maps of the populated areas in Britain, to represent facts about the British people equitably, but still to show how each area is related to others spatially. Visualization is about how to see both the detail and the whole picture (Tufte, 1990). As interest grows in high resolution mapping of populations in large areas, so the necessity of employing cartograms will increase. A traditional map of Britain that showed every ward would require an entire atlas of insets, yet a cartogram can show every ward clearly on a single page. These solutions will, however, present us with new problems. An infinite number of 'correct' population cartograms can be constructed for every grouping of any population (Sen, 1975). This is both an asset, as it allows us to choose some other properties we might wish our model to have, and a hindrance, as the superficial appearance of a cartogram of the same population will vary from one author to another. People like familiarity and that is one reason why the deficiencies of traditional map projections have been so widely ignored for so long.

One debate on the nature of cartograms is whether they should be continuous (strictly preserving topology) or not. A continuous area cartogram creates no gaps between the places represented, with all places initially neighbours remaining so. The author originally worked on this type of cartogram and considered further constraints which could be added (Dorling, 1990). For example, as shown in Plate 7, the outer boundary of the area (the perimeter) can be preserved, or the lengths of interior boundaries can be minimised. It is possible to achieve both of these aims simultaneously. As the Plate shows, maintenance of the original perimeter dramatically restricts simplification of the internal boundaries, but gives the map a familiar feel. If the shape of the perimeter is not preserved, a continuous area cartogram can then be produced which can be claimed to minimise local 'distortion'. The early work of Tobler (1961) had just such an aim and defined the term far more rigorously than it is defined here. Such a representation is very useful for some particular applications, such as the mapping of discrete incidents (for example incidents of a possibly contagious disease, see Levison and Haddon, 1965).

For mapping the general characteristics of a population, however, there is much merit in adopting the noncontinuous form of cartogram (Olson, 1976). Most areal units have relatively simple shapes in physical space which can become very complex on a continuous area cartogram. If a noncontinuous area cartogram is constructed, areal units can be represented by any desirable shapes such as circles. There are immediate benefits to be gained from this approach. For example, the area of each unit and hence its population is easier to gauge by eye. A greater benefit of this approach is realised later, as these simple shapes can be extensively manipulated to produce sophisticated pictures of population characteristics. The main disadvantage of this approach is that locally the geographical topology can be disrupted; although the algorithm to produce the cartogram can be devised to attempt to minimise the frequency and severity of this.

The algorithm the author designed (Dorling, 1991) began by positioning the areal units correctly on a land map and then applied an iterative procedure which slowly evolved a cartogram with the desired characteristics. The method used was to repel independently all areal units from each other in proportion to their population sizes (in order to give places with larger populations more room), while simultaneously applying forces of attraction in the directions of their original neighbours in proportion to their relative border lengths (in order to preserve the original topology when possible). Figure 11.1 illustrates the process of creating a noncontinuous cartogram of the counties of Britain. Places bordering the sea had a degree of inertia because part of their perimeter, being coastline, did not make up a common border. This helps to maintain prominent peninsulas and other landmarks. Thus, although the exact shape of the coastline was sacrificed, many of its well-known features could actually be retained. The algorithm has been successfully applied to create population cartograms based on as few as seventeen areal units and as many as 100 000. Figures 11.2 and 11.3 show a population cartogram and conventional map of the districts of Britain. The altered geography has been indicated on Figure 11.2 by the pattern of the lines shown connecting neighbouring areal units. To create cartograms of a few dozen units using this algorithm implemented on a home microcomputer required a few seconds, a few hundred units required a few minutes, a few thousand required a few hours, and many thousands required several days! Cartograms have been constructed of the counties, districts, constituencies, wards, census enumeration districts and several other geographical subdivisions of Britain. The algorithm can be applied to any place of any size, from the countries of the world to the buildings in a town. Its implementation is theoretically parallel in operation and in practice not necessarily limited to two dimensions.

PAINTING IN POPULATION SPACE

The layout of the cartogram produced is of interest even before we begin to

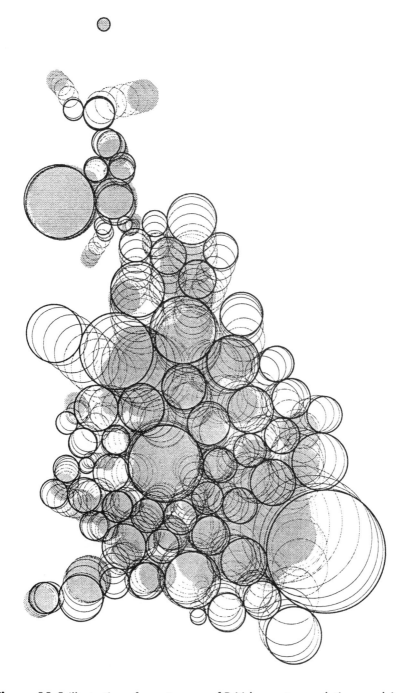

Figure 11.1 Illustration of a cartogram of British county populations evolving

Figure 11.2 Contiguity on the local authority districts population cartogram

Figure 11.3 Conventional local authority districts base map and key

use it to depict other information. A population cartogram tells us a lot about the human geography of places, how they are related to each other in a new and intriguingly unfamiliar way. We can see from Figure 11.2 that nearly half the people of Britain come under London's immediate influence and that this structure is repeated recursively since the extended Glasgow area makes up more than half of Scotland and dominates that country. Working with this visualization of the population rapidly alters the way you think about the human geography of Britain and the patterns of people's lives within it. Comparing a traditional map and cartogram of the same areal units side by side graphically illustrates how the majority of the people who live in the towns and cities are dominated by rural dwellers in the maps normally used to study them, maps which would be more appropriate for depicting the geography of sheep than people.

How the characteristics of a population can be depicted on this kind of cartogram depends on the number of areal units used or, more precisely, the final size of each unit. With over 100 000 Census Enumeration Districts, for example, each appears as a tiny dot on the screen or paper. This is large enough for the viewer to see its colour, but not its shape or small variations in its size. Nevertheless, quite sophisticated schemes can be used, from simple grey-shading to bivariate or trivariate colouring to show the distributions of one, two, or even three, population groups simultaneously. Using red, blue and yellow as primaries, mixing into purple, green and orange combinations, can produce dramatic but complex pictures such as that shown on Plate 8. The distribution of British people by their country of birth (predominantly England, Scotland or Wales) is shown by the choice of one of a possible sixty-four colours (mixtures of four levels of red, blue and yellow respectively) in each of 129 211 populated enumeration districts. The dominance of a few shades show how distinct the national divisions within Britain are, but this pattern is broken up most clearly in London where large numbers of Scots and Welsh migrants settle, colouring the western side of the capital green. Areas of white show places where large numbers of people born outside Britain have settled, for example on the east side of London, and in distinct parts of the West Midlands, Manchester, Leicester and Bradford. The equivalent map, Plate 9, swallows these people in the centres of cities and instead highlights the settlement of American airforce families in sparsely populated parts of East Anglia as white areas. The colour mixing of the Border Counties is interesting, as purple and orange are blended, but this picture tells more the story of the land than of the people. Migration to and in Britain is mainly a tale of people in cities, not of the mixing of people in remote farmland on the borders of nations, or the small scale invasion of previously unpopulated airfields.

When only 10 000 areal units are being depicted it is possible to give each a black or white border to allow them to be more distinctive and so reduce the effect that neighbouring colours have upon one another. Variation in size is easily seen and is appropriate when areal units vary a lot in size, but at this scale one cannot usefully vary shape. Figure 11.4 is a cartogram showing the distribution of unemployment in 10 444 wards in Britain at the time of the 1981 census. The important areas of high unemployment are clearly seen in Inner London, South Wales, Birmingham, Liverpool, Tyne and Wear and Glasgow. At the same time, their extent, the number of people affected, is also

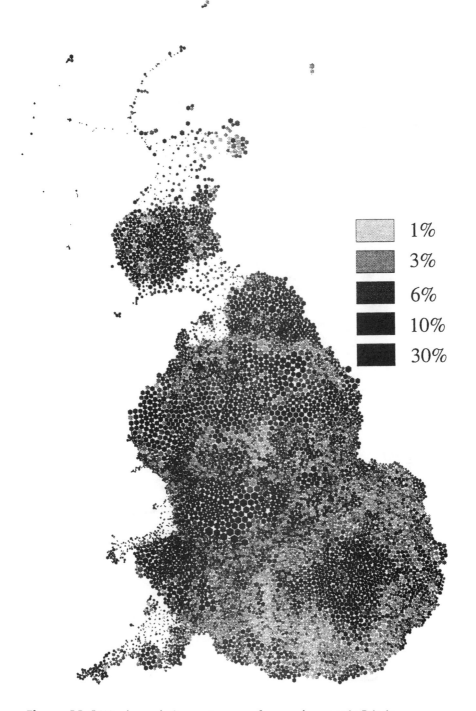

	1%
	3%
	6%
	10%
	30%

Figure 11.4 Ward population cartogram of unemployment in Britain

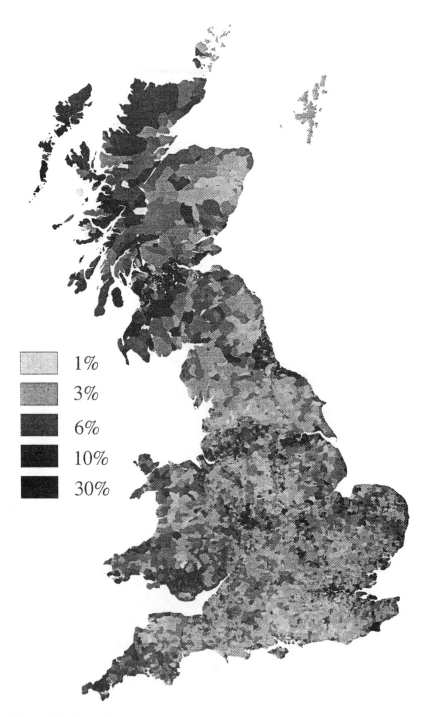

Figure 11.5 Ward map of concentration of unemployment in Britain

plain to see. Conversely, the traditional map of these same data, Figure 11.5 has a larger area than all of these coloured black in the remote northwest of Scotland, the least populous area of Britain. The major areas of unemployment now appear literally as small black spots, and the map gives the impression that most people live with little fear of losing their livelihoods. Traditional maps highlight prosperity because more affluent people tend to live in less densely populated areas.

... ELABORATING THE DESIGN

As the number of areal units is reduced to less than 1000, many more visualization opportunities present themselves. Cartographically, an advantage of using a population cartogram consisting of a mere thousand areal units is that each unit becomes large enough to alter individually, so that non-overlapping symbols called 'glyphs' (Anderson, 1989) can be put in their place. Glyphs are 'sculptured characters'; after fixing their position, colour and size we still have control over the shape and orientation of the objects which represent our cases or places. At the simplest level individual glyphs can be given a length to add an extra variable, or given direction (for example by giving the symbol an arrowhead). Symbols can be divided into two, to show the proportions of, say, men and women, or the situation before and after an election; but divisions into greater numbers often break up the spatial patterns visually. The success of such methods depends on several factors: how 'natural' we find the particular way of representing each facet; how much a group of somewhat similar glyphs creates an overall meaningful impression; and how much inherent pattern there is to the data. A picture that shows no pattern may be showing the truth, but is visually a failure.

At the simplest level the circles can be split into rings, coloured by value for two time periods to show, for example, the change of political party in parliamentary constituencies at elections. Unfortunately this schema visually fails when more rings are introduced, as the temporal pattern appears to dissect the spatial one. Trying to picture patterns that are spread across both time and space is not easy. Recalling the example of political change, suppose we were interested in the swing of the vote rather than simple change of party. The swing has two components, a direction and a magnitude and in a three-party system the direction is itself two dimensional. Upton (1991) has shown how a set of arrows can be drawn to represent the swings of votes between elections on a cartogram base. The result is most effective, even more so when you colour the arrows according to the (red/yellow/blue) mix of the votes in the constituencies to show in which political direction the swing is. Further animation of such pictures is illustrated by Dorling (1992).

Far more complex situations can be visualized. I will illustrate with some examples I have tried. The first method is to show a bar chart or population pyramid in each place. The chart is best filled in a solid colour to form visually a shape which can be quickly scanned when comparing different places or looking for regional patterns. Figure 11.6 shows the distribution of employment by eight types of industry, employee gender and status (full- or part-time) for each constituency. The area of each chart is in proportion to the working

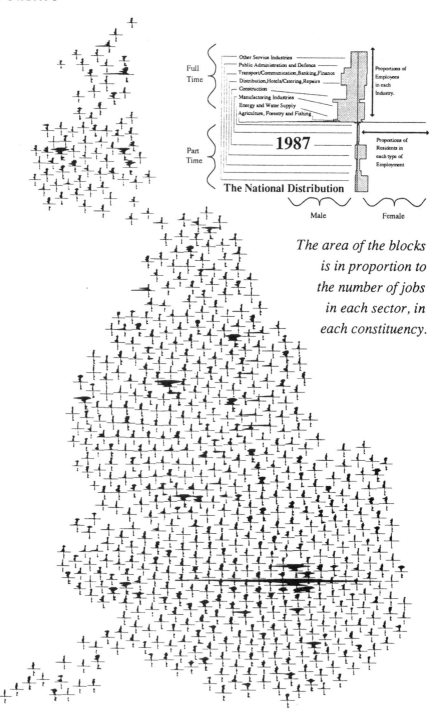

The area of the blocks
is in proportion to
the number of jobs
in each sector, in
each constituency.

Figure 11.6 The distribution of industry on a cartogram of parliamentary constituencies

Plate 1 A highway by-pass is visualized using 1:25 000 contour data converted to a TIN model. Intergraph design files for the by-pass alignment were provided by the highway authority. The wide area aerial photographs (1:20 000) were taken in summer and retinted to match the large scale (1:3 200) photographs taken along the right-of-way. When scanned the ground coverage of texture pixels was 2.5m close to the camera and 10m in the distance. In an animated fly-over the cars moved along the by-pass

Plate 2 A transmission line is visualized with landcover texture being drawn from a generic library of scanned aerial photographs. The textures are then assigned on the basis of a macGIS based landcover data set. Custom software was used to raise the Z value of forested areas, to generate vertical polygons at the forest boundary and assign a ground level captured texture to the edges

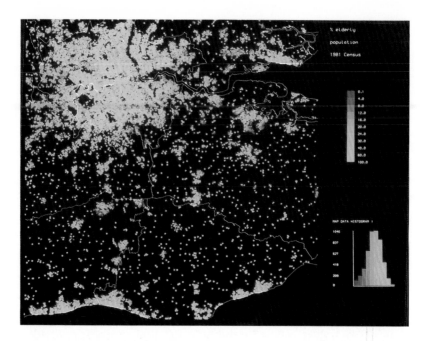

Plate 3 London and the South East: 200m cell raster image of the percentage of population aged 60 or over (1981 Census of Population)

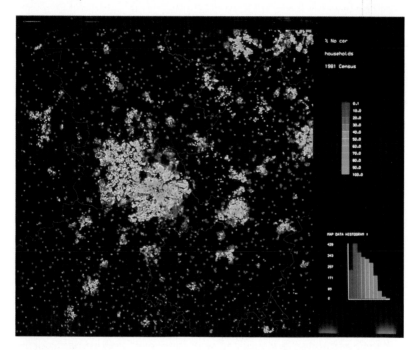

Plate 4 Birmingham/Coventry: 200m cell raster image of the percentage of non-car owning households (1981 Census of Population)

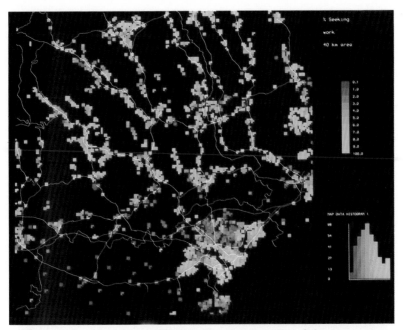

Plate 5 South Wales: 200m cell raster image of the percentage of population seeking work (1981 Census of Population)

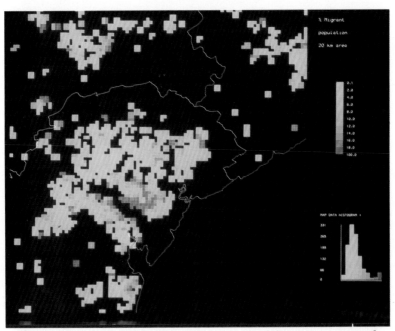

Plate 6 Cardiff: 200m cell raster image of the percentage of population with a different address one year prior to the census (1981 Census of Population)

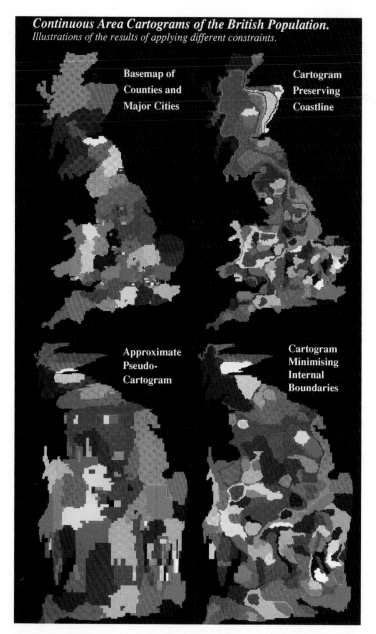

Continuous Area Cartograms of the British Population.
Illustrations of the results of applying different constraints.

Basemap of Counties and Major Cities

Cartogram Preserving Coastline

Approximate Pseudo-Cartogram

Cartogram Minimising Internal Boundaries

Plate 7 Continuous area cartograms of the British population. The same 250 counties and major cities are shown, identically coloured on each of the four images. Top left is the basemap, next to that is a continuous area population cartogram preserving the physical coastline. The second cartogram follows a suggestion by Tobler (1986) where the marginal distributions are made uniform. The final image uses the boundary of the latter, but each area is in proportion to its population, with the lengths of internal boundaries being kept to a minimum. While an interesting exercise in using cellular automata, such continuous area cartograms are difficult cartographically to embellish further

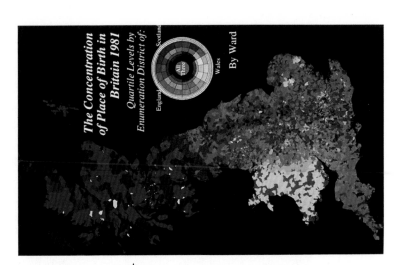

The Concentration of Place of Birth in Britain 1981

Quartile Levels by
Enumeration District of:

England   Scotland

Wales

By Ward

Plate 8 Enumeration District population cartogram showing place of birth in Britain. Each one of 129 211 populated census enumeration districts is shown coloured according to the mix of people living in it, in population space. The more English people the more coloured red whilst Scots are coloured blue and Welsh yellow. These colours mix to show the South West peninsula as generally orange, Scottish borders as purple and the western side of London green. Distinct localities can also be made out. Corby, for example, shows as a greeny blue patch of high Scots (and some Welsh) migration. West of it is Leicester (speckled white), where high numbers of people born outside of Britain have settled

Plate 9 Ward map of the concentration of place of birth in Britain. Enumeration district boundaries were not digitised for the 1981 census, and so the highest resolution conventional choropleth maps that could be drawn at was ward level comprising some 10 444 areal units. The simple regional pattern of Figure 11.2 is again evident but areas of sparse population steal our attention. White polygons in the mountains of Scotland highlight the presence of a few foreign visitors, the same in Wales and on the moors of Devon. The picture is misleading because it creates uniformity over areas where there is little, and apparent variety where few people live. The diversity in London is hidden in an area smaller than the size of a single large ward in the far north of England

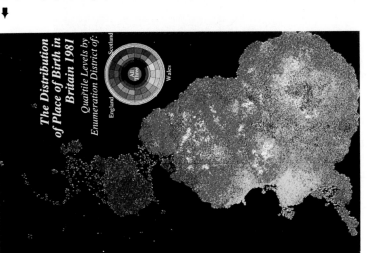

The Distribution of Place of Birth in Britain 1981

Quartile Levels by
Enumeration District of:

England   Scotland

Wales

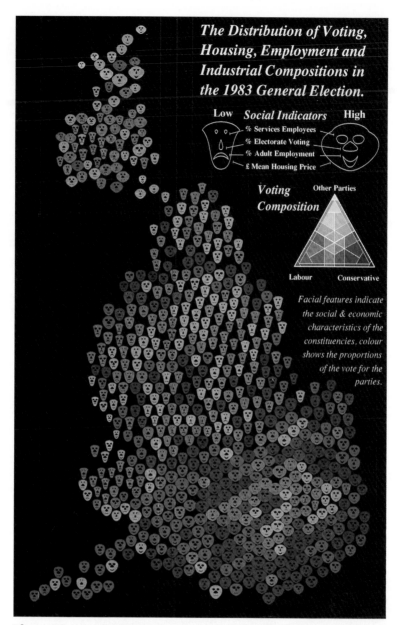

The Distribution of Voting, Housing, Employment and Industrial Compositions in the 1983 General Election.

Low **Social Indicators** High

% Services Employees
% Electorate Voting
% Adult Employment
£ Mean Housing Price

Voting Composition

Other Parties

Labour   Conservative

Facial features indicate the social & economic characteristics of the constituencies, colour shows the proportions of the vote for the parties.

Plate 10 The distributions of voting, housing, employment and industry on a population cartogram. The 633 mainland parliamentary constituencies are each represented by a face whose features express the various variables, and which is coloured by the mix of voting, drawn on an equal electorate cartogram. The patterns in this picture are very interesting and could lead to endless discussion. The 'deaths-heads' inside Glasgow city are solidly red, while the happy-faces around the capital voted strongly for the government of the day. The Welsh may not have had much employment, or expensive housing, but they still turned out to vote in large numbers. This technique is particularly good for identifying exceptions, faces which do not fit in with the crowd

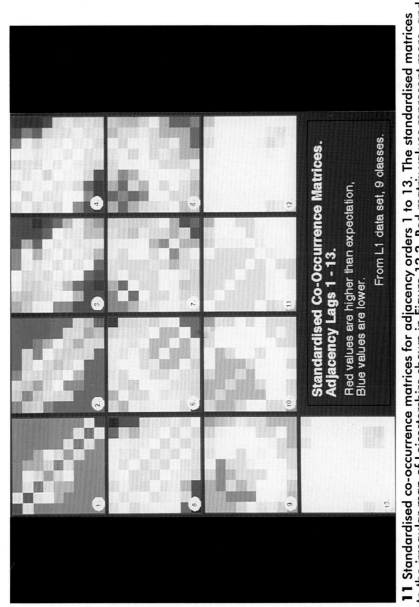

**Standardised Co-Occurrence Matrices.
Adjacency Lags 1 - 13.**

Red values are higher than expectation,
Blue values are lower.

From L1 data set, 9 classes.

Plate 11 Standardised co-occurrence matrices for adjacency orders 1 to 13. The standardised matrices relate to the irregular map of Leicestershire shown in Figure 12.2. The standardised matrices blue values fewer, occurrences than the expected value. The counts matrices (see Figure 12.2) are standardised by the mean and standard deviation of the expected number of co-occurrences of each class under the overall constraints of non-free sampling conditions

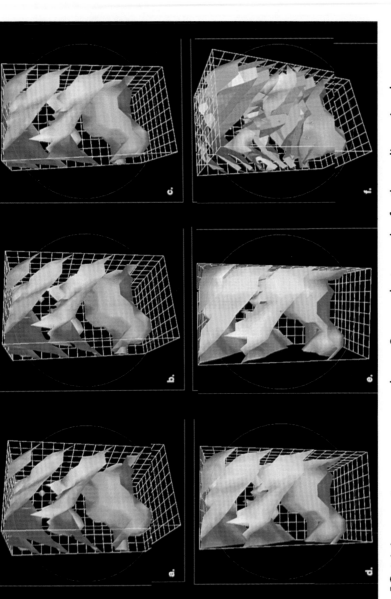

Plate 12 A joins count co-occurrence volume. Screen photograph of a three-dimensional co-occurrence volume rendered using interactive ViSC software (SGI's Explorer). The horizontal axes relate to data classes, the vertical to lags in adjacency order. Order 1, direct adjacency, is represented by the top plane. Successive horizontal slices represent other adjacency orders. The co-occurrence matrix data values are plotted at the nodes of the referencing mesh and the data relate to the map shown in Figure 12.2. Here iso-surfaces have been created at values of 0 (pink) and 40 (yellow) to visualize the distribution of data within the volume. The sloping yellow surface reflects the way in which the map is autocorrelated at low lags

population. Because of regional patterns, the industrial structure can be assimilated visually from several hundred pyramids, spaced out on the cartogram (although interactive zooming was useful to inspect such detailed charts).

Second, the distribution of house prices has been depicted by using a glyph shaped like a tree. The tree branches into different types of housing divided by features such as number of bedrooms, bathrooms, heating and detachment. The width of each branch shows the number of sales of that type of house, the length is in proportion to the average price. It follows that the total depicted branch area gives financial turnover for housing in that spatial area. The result, shown as Figure 11.7, looks like a wood with trees of different species, sometimes occurring in identifiable clumps of particular sizes and shapes. Again, because there is pattern to the distribution, a meaningful picture is created.

An especially fascinating glyph is one based on human faces, first drawn by Chernoff (1973; 1978). Facial expressions, it is argued, are one of the visual images we are best equipped to decipher. We naturally combine their features to interpret moods such as happy or sad, sly or simple. What is more, we can easily compare faces and pick out similarities and exceptions in a crowd. Faces maintain a basic structure within which even slight variation often holds meaning. The original Chernoff faces were designed to portray up to eighteen variables and statisticians have subsequently extended this to thirty-six (Flury and Riedwyl, 1981), but psychological research suggests that the permutation of so many different variables to features resulted in staggering differences in interpretation (Chernoff and Rizvi, 1975). This is hardly surprising as many combinations of variables in the original can visually cancel each other out. It is difficult to see the slant of an eye when its size was reduced to a point! Here, I have been somewhat less ambitious and employed only five features and variables to look at some of the factors behind voting swings across Britain. Figure 11.8 shows the construction lines of the Bezier curves used to create the graphics.

Parliamentary constituencies were drawn as heads, with variation in area to show the total number of voters, width to show house price (fat cheeks when expensive!) mouth style to show employment rate (a smile when high), nose size for electoral turnout, and eyes to show industrial structure (large and low when dominated by younger industries). The colour of the face represented the actual voting patterns. The study was initially meant as a 'tongue-in-cheek' exercise but the result, shown in Plate 10, was revealing. The inner–outer city and north–south divides are in many aspects clear, but the difficulty of drawing precise lines between the regions and around the cities is also apparent. Specific individual faces can be identified which did not 'fit in' (as when plotting swings some arrows did not 'go with the flow'). Strong local trends and relatively sharp divisions are the clearest messages of this visualization. The faces were also used to show change over time, coloured and shaped by the swings in the votes, change in house prices, unemployment increase and so on. Thus the facial expressions become populations' 'reactions' to a changing situation, their colours perhaps indicating a political response by the electorate to economic change. Further psychological work would be valuable to determine how much the apparent increase in insight gained through using these particular glyphs might be outweighed by the possible ambiguities introduced through such an emotive visualization.

Figure 11.7 The distribution of owner occupied housing on a population cartogram

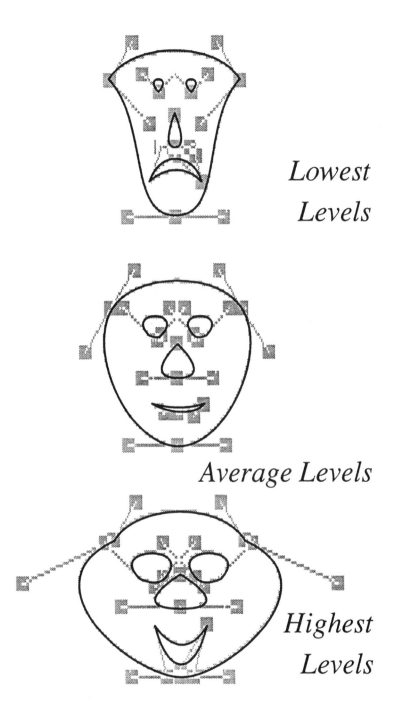

Lowest
Levels

Average Levels

Highest
Levels

Figure 11.8 The mechanics of constructing glyphs of faces

... CONCLUSION

The argument for the use of cartograms in visualizing human geography has been developed from outlining the geographic disadvantages of not doing so, to the cartographic advantages of using an uncluttered map on which the people actually fit. The arguments against using cartograms have not been put, but it should not be difficult for the reader to begin to list them, so strongly is our present geographical understanding based on the topographic base. Recall that this was originally designed to help sailors over oceans, not to find people in cities. You know where things are on an ordinary map. However, there is nothing fundamentally new in using area cartograms. You already use road maps where the roads are enormous compared to reality and underground maps whose shape is drawn to give a 'clearer picture'. Yet throughout cartographic history people have been developing more appropriate projections for mapping their times and places. These have often met with resistance if they offered a new perspective on the world. Whatever its absolute merits, the so-called Peters projection received considerable criticism for doing just that (Porter and Voxland, 1986); you can expect people not to like the shape of their world being changed!

A number of alternatives to cartograms have been suggested in the literature (and are illustrated elsewhere in this section of the book), but often result in greater disadvantages. A population density surface can be constructed upon an equal land area map (in effect an equal volume population cartogram) to expose those hidden in cities as 'mountains' on this surface, but our ability to gauge volume is poor, and should be avoided when possible. There are also the well-known problems in visualizing surfaces related to orientation, shadow, perspective, hidden sections and so on. There is a most complex relationship between what is actually seen on the paper and what the picture is intending to show. If we are to resort to 3D then drawing a surface upon a cartogram base allows for far more sophisticated and straightforward analyses, for example, a landscape of house prices coloured by unemployment over population space where height is price.

A second substitution sometimes mentioned is to use the interactive animation outlined in Chapters 13 and 14 to focus on the small highly populated areas. Again the usual complaints with insets can be made: we want to be able to see the detail and the whole simultaneously. In contrast, zooming in on a highly detailed area cartogram can allow you to see that patterns perceived at the large (human) scale are often repeated at the small scale (for instance rich and poor areas can be found at each scale). Investigate the distribution within the population, not the distribution of the population.

Cartograms can be seen today as a kind of artificial reality which we deliberately construct to obtain knowledge. They allow us to optimize the visualization of a chosen body of information, and with population cartograms we wish to give every person equal prominence in our picture. If we are to understand the spatial structure of society we must find effective ways of envisioning it. We have to open up our maps to show us all the people, not hide the majority in tiny dots on an agricultural map. Then we can employ a whole battery of techniques – shade, colour, shape, symbols, statistical analysis, even surfaces and animation if necessary – to depict and examine the information we have about the people who live in these tiny places. Cartograms should not be seen as just another option in a cartographic toolbox, but as a fundamental necessity in the *just* mapping of spatial social structure.

... ACKNOWLEDGEMENTS

Thanks are due to Zhilin Li, Chris Brunsdon and David Dorling who commented on an earlier draft of this chapter. Many of the participants of the visualization workshop from which this book evolved also helped shape the final version of this text.

... REFERENCES

Anderson G C (1989) Images worth thousands of bits of data. *The Scientist* 3(3): 1, 16–17

Angel S, Hyman G M (1972) Transformations and geographic theory. *Geographical Analysis* 4: 350–367

Beniger J R (1976) Science's 'unwritten' history: the development of quantitative and statistical graphics. *Abstracts of the Annual Meeting of the American Sociological Association*, New York

Bunge W (1966) *Theoretical geography*. Lund Studies in Geography, Series C, 1: 285pp

Census Research Unit (1980) *People in Britain – a census atlas*, HMSO, London

Chernoff H (1973) The use of faces to represent points in k-dimensional space graphically. *Journal of the American Statistical Association* 68(342): 361–368

Chernoff H (1978) Graphical representations as a discipline. In: Wang P C C (ed.) *Graphical Representation of Multivariate Data*, Academic Press, New York

Chernoff H, Rizvi M H (1975) Effect on classification error of random permutations of features in representing multivariate data by faces. *Journal of the American Statistical Association* 70(351): 548–554

Coulson M R (1977) Political truth and the graphic image. *The Canadian Cartographer* 14(20): 101–111

Craig J (1976) 1971 census grid squares. *Population Trends* 6: 14–15

Dorling D (1990) A cartogram for visualization. *North East Regional Research Laboratory Research* Reports 90/4, University of Newcastle upon Tyne, Newcastle upon Tyne UK

Dorling D (1991) *The visualization of spatial social structure*, Unpublished PhD thesis, University of Newcastle upon Tyne, Newcastle upon Tyne UK

Dorling D (1992) Stretching space and splicing time: from cartographic animation to interactive visualization. *Cartography and Geographic Information Systems* 19(4): 215–227

Dougenik J A, Chrisman N R, Niemeyer D R (1985) An algorithm to construct continuous area cartograms. *Professional Geographer* 37(1): 75–81

Eastman J R, Nelson W, Shields G (1981) Production considerations in isodensity mapping. *Cartographica* 18(1), 24–30

Flury B, Riedwyl H (1981) Graphical representation of multivariate data by means of asymmetrical faces. *Journal of the American Statistical Association* 76(376): 757–765

Harley J B (1990) Cartography, ethics and social theory. *Cartographica* 27(2): 1–27

Hollingsworth T H (1964) The political colour of Britain by numbers of voters. *The Times*, 19 October, p. 18

Hollingsworth T H (1966) The political colour of Britain by winning parties. *The Times*, 4 April, p. 8

Levison M E, Haddon W (1965) The area adjusted map: an epidemiological device. *Public Health Reports* 80(1): 55–59

Olson J M (1976) Noncontiguous area cartograms. *Professional Geographer* 28: 371–380

Porter P, Voxland P (1986) Distortion in maps: the Peters projection and other devilments. *Focus* 36: 22–30

Raisz E (1934) The rectangular statistical cartogram. *Geographical Review* 24: 292–296

Raisz E (1936) Rectangular statistical cartogram of the world. *Journal of Geography* 35: 8–10

Rhind D (1983) Mapping census data. In: Rhind D (ed.) *A Census User's Handbook*, Methuen, London

Sen A K (1975) A theorem related to cartograms. *American Mathematics Monthly* 82: 382–385

Skoda L, Robertson J C (1972) Isodemographic map of Canada. *Geographical Papers 50*, Department of the Environment, Ottawa, Canada

Tobler W R (1961) *Map transformations of geographic space*. Unpublished PhD thesis, Department of Geography, University of Washington, Washington USA

Tobler W R (1963) Geographic area and map projections. *Geographical Review* 53: 59–78

Tobler W R (1976) Cartograms and cartosplines. *Proceedings of the Workshop on Automated Cartography and Epidemiology*, National Center for Health Statistics, US Department of Health, pp. 53–58

Tobler W R (1985) Interactive construction of continuous cartograms. *Proceedings of the 7th International Symposium on Computer-Assisted Cartography*, Washington DC, p. 525

Tobler W R (1986) Pseudo-cartograms. *The American Cartographer* 13(1): 43–50.

Tufte E R (1990) *Envisioning Information*, Graphics Press, Cheshire, Connecticut

Upton G J G (1991) Displaying election results, *Political Geography Quarterly* 10(3): 200–220

12 AREA-VALUE DATA: NEW VISUAL EMPHASES AND REPRESENTATIONS

J. DYKES

... VISUALIZATION AND MAPS

The visual depiction of data has various purposes. Ideation, the quest for understanding, may be aided by a visual representation of a dataset. Once knowledge is acquired it can be disseminated to a mass audience through simpler images. Graphic communication as a means of explanation can be traced through the centuries (Ferguson, 1977) and cartographic representations of multidimensional data have populated both ends of the user continuum introduced in Chapter 7. However, the effort required to produce them has traditionally meant that single visual representations have had conflicting functions as both accurate data inventories for ideation and as vehicles for conveying specific messages (Board and Taylor, 1977).

Because technology now allows large numbers of maps to be generated at speed, map objectives are changing. In addition, recent developments in computer graphics (McCormick et al., 1987) and an escalation in data volumes have given rise to Visualization in Scientific Computing (ViSC) in which scientists use interactive display to extract ideas from masses of data. This means that previous disdain for the visual depiction of data in favour of supposedly 'objective' analytical approaches is being superseded by a recognition that human vision is itself a powerful scientific tool (MacEachren and Ganter, 1990).

... THE CHANGING ROLE OF THE STATIC MAP

There is, however, a role for the static map. Traditional maps have been the basis for visual spatial analysis, so that attempts have been made to reproduce spatial datasets 'accurately'. This usually involves an adherence to the recorded statistical values. With the advent of digital data storage and interactive computer mapping this is less necessary and static maps can be used to model aspects of the data in a revealing manner for both ideation and communication. Even when analysing data interactively the spatial scientist may operate by altering visual variables to illustrate and highlight patterns in an exploratory manner until patterns are uncovered. The production of the resultant static map is thus a creative process, in which a model of reality is formed by the application of a series of processes to achieve a particular goal. With a declining need for raw data values to be tabulated graphically, the mapping of aspects of data is increasingly important. Furthermore, while McCormick's (et al., 1987, p. 6) reminder that 'one cannot publish interactive systems in a journal' may be overcome in technological terms, not everyone wants, or has the capacity, to process their own data. Further, as Gilmartin (1982) shows, static maps and iconic memory are more resistant to fading than other forms of memory.

This chapter responds to these technological advances and their implications by suggesting some new visual representations for area valued data of the type normally shown by choroplethic maps. It concerns visualization at two levels. First, it is suggested that a static area value map can be classified so that spatial association is maximised. This process has potential benefits at both ends of the user continuum. Second, it is then demonstrated that ViSC technology can be applied to the static map in order to explore and extract information from areally mapped classified datasets.

... CLASSIFICATION TO EMPHASISE SPATIAL RELATIONSHIPS

In introducing the idea of a 'classless' choropleth map, Tobler (1973) argued that the most useful representation would have as many grades of shading as data values. Others, notably Dobson (1973), have countered this by noting that classless choropleths do not model or provide insight. The resolution of this debate may well depend on the intended location of any given map along the user continuum. However, in communication terms regrouping data items into classes decreases the communicative bandwidth and reduces the potential for individual vagaries of interpretation. These reasons for classifying are summarised by Muehrcke (1990, quoted in Buttenfield and MacKaness, 1992) who notes that 'it is abstraction, not realism that give maps their power'.

Once free from the constraints of having to be accurate data inventories, maps can be used to display spatial pattern at the expense of other functions. This use of maps to model pattern is termed 'map analysis' by Muehrcke (1978) and 'intermediate map use' by Bertin (1981). The incorporation of spatial associations as classification parameters instead of statistical goals such as 'map accuracy' (Jenks and Caspall, 1971) which is explored in this chapter is one example. Traditional classification techniques categorise a series of values in a single dimension and then map them, yet if space per se is given sufficient status in solving a geographical problem, maps may be produced that communicate spatial distributions more effectively (Muehrcke, 1972; Gatrell, 1974). The goal of internally homogenous classes (Jenks and Caspall, 1971) is thus subordinated to the communication objective (Monmonier, 1974). This is an important recognition, and one that has frequently been overlooked. For example, Evans (1977) acknowledges 'contiguity biased' class selection, but does not advocate it, and Egbert and Slocum (1992) have no spatial association classifier in their interactive choropleth visualizer.

Several features of the choropleth negate against statistical accuracy, and so add to the argument for the inclusion of a spatial pattern component in classification. For example the data are masked by aggregation and the effects of zone size and shape as filters and the imposition of unnatural boundaries has a considerable effect on aggregate values.

Finally, the production of a spatial classifier satisfies a further demand. Dawsey (1990) refers to the need for aesthetically pleasing maps without continuous user interaction, and Weibel and Buttenfield (1992) suggest that pre-specified map designs are supplied for non-proficient users of interactive systems. The incorporation of a spatial classification algorithm may help fulfil these requirements.

... THE APPLICATION OF AN AUTOCORRELATION COEFFICIENT TO AREA VALUE MAPS

In order to quantify the spatial arrangement of areal values in a classified choropleth map, a spatial statistic is required. Numerous map complexity coefficients have been applied to area value data (see Muehrcke, 1973; Olson, 1974; Muller, 1976) but the most rigorous those based on spatial autocorrelation. According to Goodchild (1986, p. 5): '... if one were to summarise a spatial distribution of unequal attributes in a single statistic, one would in all likelihood choose a spatial autocorrelation index.'

In order to examine the potential for varying map autocorrelation by class delimitation on choropleth maps, Dykes (1991) used Moran's I. This is calculated as the weighted sum of the product of the data centred on their expected value, standardised by their variance, and normalised for the total sum of the weights (Cliff and Ord, 1981). A $1/d^2$ inverse distance squared weight function was used. A series of data distributions with varying degrees of spatial dependence and different statistical characteristics was used with a regular 10 by 10 grid and an irregular set of polygons (the 166 Wards of Leicestershire) to evaluate the effects of choropleth classification on autocorrelation. To keep the classes realistic thirteen classification algorithms were implemented, and each set of zones was classified into three, five, seven and nine classes for the variety of distributions.

The regular grid illustrated that autocorrelation can be altered by varying class delimitation strategies but the range of effects was data specific. Random distributions, or those that varied in a spatially uniform manner, displayed less fluctuation in autocorrelation when class boundaries were altered than those which combined a spatial trend with a random element to produce a complex, but autocorrelated, surface. Informal analysis of the coefficients illustrated that for a simple grid, the visually interpreted association matches relatively well with Moran's I. This can be seen by inspection of Figure 12.1(a)–(c).

Unfortunately, the coefficients returned for the irregularly zoned datasets proved impossible to estimate visually (see Figure 12.1(d)–(f)). This result illustrates a central problem in ViSC methodology, which seeks to make use of our pattern recognition capabilities and the proficiency of the eye to absorb large amounts of information very rapidly. Lloyd and Steinke (1976), Muller (1976) and Peterson (1985) all suggest that individuals have very different bases for identifying pattern in irregularly zoned choropleths. Openshaw (1992) notes that the eye/brain is easily misled and that spatial patterns that actually occur are often far too complex for visual recognition and interpretation.

105

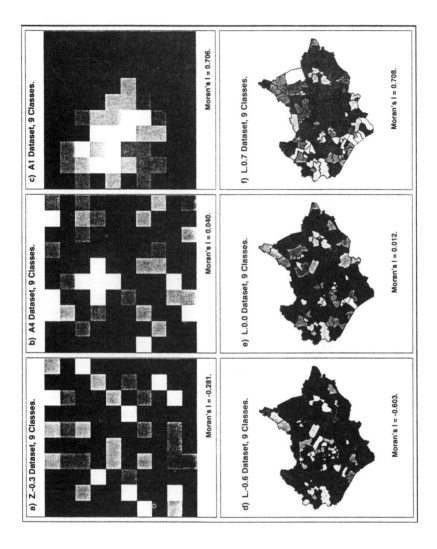

a) Z.-0.3 Dataset, 9 Classes. Moran's I = -0.281.

b) A4 Dataset, 9 Classes. Moran's I = 0.040.

c) A1 Dataset, 9 Classes. Moran's I = 0.706.

d) L.-0.6 Dataset, 9 Classes. Moran's I = -0.603.

e) L.0.0 Dataset, 9 Classes. Moran's I = 0.012.

f) L.0.7 Dataset, 9 Classes. Moran's I = 0.708.

Figure 12.1 Maps depicting simulated statistical surfaces which show a range of values of Moran's I.

It would seem that the way in which an autocorrelation coefficient biases the outcome toward variation between very close neighbours runs directly counter to the way in which mapped trends are recognised through areas of contiguous shaded blocks. For example, an homogenous series of values surrounding an extreme deviation will produce a low (not autocorrelated) coefficient despite the evident homogeneity in a high proportion of the mapped space. This effect is especially acute where the values represent small zones, as the $1/d^2$ weight function is very sensitive to exceedingly close centroids, whatever the values of the visually dominant larger zones. As well as these failures of the coefficient to match perceived patterns, the coefficients are reduced in utility by their employment of simple Euclidean geometry. Using physical distances between centroids generated from irregular zones is naive both in terms of map pattern quantification and the measurement of spatial processes.

These findings suggest that Moran's I and other distance-based autocorrelation coefficients, do not describe pattern parameters that can be used successfully as a basis for illustrating visually recognisable data trends when classifying irregularly zoned areal datasets. This is an important conclusion because many of the maps that have been used previously to measure complexity involve fewer zones of substantially less size variation than those used in this analysis (see, for example Jenks and Caspall, 1971; Muller, 1975; Lloyd and Steinke, 1976). By studying irregular zones that approach a regular lattice, the influence of the weight function may not have been fully appreciated. This highlights the need for continued research into finding a function for extracting the spatial information contained in a complex area valued dataset.

Further restrictions of the spatial autocorrelation index involve uncertainty surrounding the definition of a standard weight function meaning that Moran's I is no different from many other weighted spatial statistics in that it may be manipulated to 'prove' that a dataset behaves in a particular manner. An additional constraint with this technique concerns its inability to examine the spatial relationships for neighbours other than those immediately adjacent, which is desirable when examining complexity amongst areal values (Gatrell, 1974).

These difficulties in relating a measurable statistic to a complex map have long been recognised by thematic cartographers. Muehrcke (1972, p. 53) states that:

> visual map reading does not permit the recognition or differentiation of subtleties in configuration, and rapidly deteriorates in quality as control gets sparse, pattern grows complex and the number of variables increases. For these and other reasons we suspect that the results obtained from visual map reading are often imprecise, inconsistent, or more apparent than real. If these deficiencies were, in fact, real, they would distract significantly from the use of maps in scientific enquiry, especially if they went unrecognised.

This conflict between statistically measured and visually perceived trends has an important bearing upon the visualization of irregular area value data. If sophisticated autocorrelation indices do not reflect visually perceived trends, then the use of such statistics as a basis for map classification that emphasises a pattern that can be communicated effectively are doubtful. Furthermore, if measured patterns cannot be perceived questions arise concerning the appropriateness of the choropleth for communicating geographical pattern and ideation. Single statistics do not provide adequate information from the mapped data and yet an irregular choropleth is too complex for visual analysis. Consequently there is a need for a function that does reflect visual trends so that effective, spatially classified maps can be produced and a method of displaying area value data that facilitates the analysis of map pattern can be developed.

...ANALYSING AUTOCORRELATION BY ABSTRACTING SPATIAL ASSOCIATIONS

The complexity of zone shapes on the irregular map, the restrictions associated with the distance modelling function, and the way in which gridded data prove much more amenable to visual analysis suggest that a non-metric representation that retains the topology of the original zones may assist in the interpretation of mapped trends. At the cost of the metric information, portraying map pattern in regular non-metric space may well illustrate trend and help to match pattern with statistic.

The nature of area value data strengthens this conviction. Through their method of capture, area value data exhibit an intrinsic paucity of distance-related information, and attempts to compensate for this such as centroid calculation, $1/d^2$ modelling, and common boundary length coefficients are inadequate. Having noted the limitations of metric relations, the only remaining spatial parameter is adjacency and it is proposed that a wealth of spatial information is contained within the adjacency relations network. From this it is suggested that zone adjacency is the essential feature of the choropleth map, and should form the basis of a map autocorrelation function.

A simple single joins count statistic measures map connectedness and may provide such a function, but it fails to account for data ordinality in multinomial maps and is thus unsatisfactory. In the current research a multinomial adjacency-based autocorrelation function is obtained by borrowing a technique used originally in image processing. Haralick (1979; et al., 1973) assumed that the information contained in an image can be summarised by the adjacency relationship which grey values have to one another. They compiled 'spatial-dependence probability-distribution matrices' for a given set of grey values within an image. This information is adequately specified by the matrices which record spatial association, whereby the axes relate to attribute class and values relate to the number of occurrences of each combination of classes as neighbours. Although Haralick et al. (1973) produced matrices relating to particular directions and adjacency lags in a raster image, similar co-occurrence matrices can be constructed from choropleth maps if classes are used instead of grey tones, and counts are recorded between each combination of classes. Matrices can thus be extracted for adjacency orders from one (immediate neighbours) up to and including the diameter of the map's joins network. This method of characterising a map's pattern is particularly suited to communicating negative and lagged spatial autocorrelation patterns, as the natural emphasis is on extracting information regarding relationships between proximal zones from the choropleth map, and negative or lagged autocorrelation are difficult to detect visually unless they are extreme. The technique combines the various levels of association at lags, a feature which is desirable, but lacking, in most statistical analyses of autocorrelation (Gatrell, 1974). It is analogous to the spatial correlogram which describes the autocorrelation structure (Haggett et al., 1977; Oden 1984).

Such co-occurrence matrices have previously been created from regular binomial maps – Gatrell (1974) termed his 2 by 2 matrix a 'Frequency Matrix of Joins' – but have not previously been subjected to any sophisticated analysis. Haralick suggested fourteen textural features which can be extracted from the matrices, through which the clustering of counts can be analysed to describe the base map associations. Although many of these coefficients are correlated, 'homogeneity' and 'contrast' appear to be particularly useful.

The usefulness of this co-occurrence matrix abstraction is best illustrated when, in accordance with the ViSC paradigm, the counts data are presented cartographically as a regular grid. Values can be mapped either as raw counts as in Figure 12.2, or as standard scores relating them to the number of joins expected from the map, if the data values were distributed randomly, as shown in Plate 11. Although abstract in that the mapping does not relate directly to real world spatial distributions, this ordering of the information contained in a classified areal dataset is analogous to the use of Chernhoff Faces (see Chapter 11). Both are mappings that attempt to retain familiarity in order to aid visualization.

ViSC techniques have been developed to help scientists extract ideas from masses of multidimensional data. They are specifically directed at those who have a deep knowledge of what they are looking at. In the present work, the expert user can take advantage of visual techniques by sequentially layering matrices for each order of adjacency to create a three-dimensional representation of the data, or 'co-occurrence volume' such as that visualized in Plate 12. By extracting contiguity characteristics and making value arrangement (Haining et al., 1983) the subject of an abstract graphic, many of the benefits associated with the ViSC methodology can be applied to the analysis of classified areal datasets. Horizontal slices through the cube illustrate occurrences of each class combination at any order, and vertical slices provide visual insight into the occurrence of a single class with all other classes at each order of adjacency. Features such as variable iso-surface rendering and moving plane probes allow the volume to be interrogated interactively.

Contrary to the majority of visualization applications which reduce multidimensional data to fewer dimensions, this method actually displays the complexities of a classified areal dataset in a higher dimensionality than they were recorded. Yet it complies with basic spirit of ViSC in that it allows us to see the unseen in order to understand the nature of spatial data (Robertson, 1988a; Buttenfield and MacKaness, 1992). The use of a third dimension allows data to be displayed on a regular lattice, in a visually manageable form and with the emphasis on extreme values lessened. The complicating impacts of zone shape and edges are removed and the complex calculations that model metric spatial associations become unnecessary.

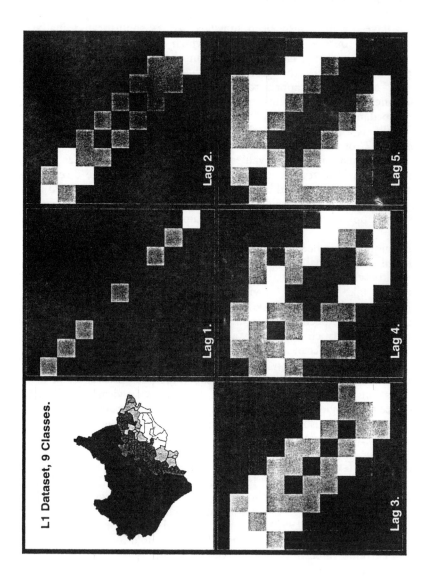

Figure 12.2 Irregular map and associated joins count co-occurrence matrices

... CURRENT RESEARCH: FAMILIARISATION WITH THE VOLUME

Research is currently progressing to explore the value of these representations, particularly the way in which their characteristics can be related to the choropleth from which they were derived. Questions currently under consideration include the following:

- Do certain types of base map produce similar or predictable coefficients, matrices and volumes?
- Do changes in the base map (through class delimitation, zone alteration, etc.) have predictable effects?
- Can users relate changes in the abstract graphics and statistics to the base map?
- Do certain mathematical functions, or visual features of the abstractions, relate to particular spatial relationships and associations?

Ultimately, in terms of the communication objective, there is a need to answer the following:

- Which of the mathematical functions and/or features of the visual abstractions relate to the aspects of the choropleth that affect the portrayal of map pattern to the map reader?
- Can rules be drawn that specify these relationships?

If the answer to the latter question is positive then these rules can be tested, and relevant functions included as variables to maximise when classifying choropleths.

... FURTHER RESEARCH: DATA EXPLORATION

At the expert end of the user continuum, co-occurrence volumes may be used to give insight into the adjacency relationships between data classes in area value maps. Visual and statistical analysis of further map parameters using the techniques outlined above will add to the understanding of the abstraction, and its representation of spatial distributions. The effects of areally aggregating continuous surface data may be explored through co-occurrence volumes, by analysing controlled datasets and varying spatial characteristics. This will provide insight into the association between agglomeration and autocorrelation, and may facilitate Muller's (1992) objective of aggregation based on autocorrelation rather than statistical classifications. Proposed experimentation involves the following:

- Imposing different zones on data surfaces and comparing the alternative volumes that result. These may be random, agglomerated or real (for example the Leicestershire data can be mapped as enumeration districts, parishes, travel to work areas, unit postcodes, etc.).
- Analysing the effects of map class delimitation strategies on spatial pattern.
- Changing the number of mapped classes to alter the resolution of the grid making up the volume.
- Analysing source map subregions, and their relationship with the overall volume. Volumetric addition for subregions will not result in the total

regional matrix. Analysing subregions in this manner may aid in the extraction of significant features.

This work may result in rules upon which decisions can be made in choropleth classification. Ultimately, the analysis could be conducted interactively by incorporating three-dimensional co-occurrence volumes as features of 2D area-value data ViSC systems in which they will be manipulated and updated in response to base map parameters being altered.

These methods of analysing spatial association between irregular zones from complex classified area value datasets result from the need for a suitable function that measures perceived spatial pattern from static maps. The application of a spatial classifier has both 'communication' and 'ideation' implications. Maps can be classified so that a simple pattern is represented with the communication objective paramount, or a spatial model can be produced that maximises the autocorrelation of the underlying spatial processes. The methods are also in the tradition of Openshaw's (1992) definition of descriptive, rather than inferential, statistics and exploratory geographical analysis tools to help users find and describe patterns and relationships. In effect, the visual representations and supportive statistics are analogous to the histogram and its descriptive statistics which have been used to characterise and classify data up until now. The degree of insight attainable will only become apparent with familiarisation, and the extension of the technique to larger and more complex datasets.

Potential applications and benefits include:

- An improved portrayal and analysis of spatial pattern in area-value data.
- Enhanced visual judgement of spatial relationships among such data.
- Map characterisation and comparison independent of zone shapes or numbers.
- Examination of the effects of spatial aggregation and zone agglomeration on the representation of a continuous surface as a series of linked zones.
- More information upon which to base decisions concerning delimitation of zone boundaries and attribute data classes.
- Rules or guidelines potentially linkable to intelligent knowledge-based systems. For example, goals for iterative class delimitation, automatic map characterisation, automatic zone agglomeration, with consideration of the effect upon spatial organisation.

The techniques take advantage of the power of ViSC technology and conflict with the traditional role of the cartographer (who reduces multidimensional data into two dimensions), by expanding the dimensionality of geographic data, or aspects of them, to fill virtual three-dimensional spaces so that they can be analysed visually. Bishop (Chapter 8) provides a 'realistic' example, and other uses of such spaces could be in visualizing multivariate scatter clouds, providing an extra dimension for multidimensional scaling solutions or plotting adjacent values against metric distance for an unclassed choropleth. It is suggested that plotting data in such spaces and using statistical surfaces and probes to analyse them has huge potential in spatial science.

... REFERENCES

Bertin (1981) *Graphics and Graphic Information Processing*, Walter de Gruyter, Berlin

Board C, Taylor R M (1977) Perception and maps: human factors in map design and interpretation. *Transactions of the Institute of British Geographers* 2(1): 19–36

Buttenfield B P, MacKaness W A (1992) Visualization. In: Maguire D J, Goodchild M F, Rhind D W (eds) *Geographical Information Systems: Principles and Applications*, pp. 19–36

Chou Y-H (1991) Map Resolution and Autocorrelation. *Geographical Analysis* 23(3): 228–245

Cliff A D, Ord J K (1981) *Spatial Processes: Models and Applications*, Pion, London

Dawsey C B III (1990) Algorithms for uniform range interval classification. *Cartographica* 27(1): 46–53

Dobson M W (1973) Choropleth maps without class intervals? A comment. *Geographical Analysis* 5: 358–360

Dykes J A (1991) *An Investigation Into the Effects of Classification Upon Spatial Autocorrelation in an Attempt to Maximise the 'Communication Effectiveness' of the Choropleth Map*, Unpublished MSc Thesis, University of Leicester UK

Egbert S C, Slocum T A (1992) EXPLOREMAP: an exploration system for choropleth maps. *Annals, Association of American Geographers* 82(2): 275–288

Evans I S (1977) The selection of class intervals. *Transactions, Institute of British Geographers* 2(1): 98–124

Ferguson E S (1977) The mind's eye: nonverbal thought in technology. *Science* 197(4306): 827–836

Gatrell A C (1974) *Complexity and Redundancy in Choropleth Maps*, Unpublished MSc thesis, Pennsylvania State University USA

Gilmartin P P (1982) The instructional efficacy of maps in geographic text. *Journal of Geography* 4: 145–150

Goodchild M F (1986) *Spatial Autocorrelation*. Concepts and Techniques in Modern Geography (CATMOG) 47, Geo Books, Norwich UK

Haggett P, Cliff A D, Frey A (1977) *Locational Analysis in Human Geography* 2nd edn, Edward Arnold, London

Haining R, Griffith D A, Bennett R (1983) Simulating two-dimensional autocorrelated surfaces. *Geographical Analysis* 15(3): 247–253.

Haralick R M, Shanmugam K, Dinstein I (1973) Textural features for image classification. *IEEE Transactions on Systems, Man and Cybernetics* SMC-3(6): 610–621

Haralick R M (1979) Statistical and structural approaches to texture. *Proceedings of the IEEE* 67(5): 786–804

Jenks G F, Caspall F C (1971) Error on choroplethic maps: definition, measurement and reduction. *Annals, Association of American Geographers* 61: 217–244

Lloyd R, Steinke T (1976) The decision making process for judging the similarity of choropleth maps. *The American Cartographer* 3(2): 177–184

MacEachren A M, Ganter J H (1990) A pattern identification approach to cartographic visualization. *Cartographica* 27(2): 64–81

McCormick B H, DeFanti T A, Brown M D (eds) (1987) Visualization in scientific computing. Special issue ACM SIGGRAPH *Computer Graphics* 21(6)

Monmonier M S (1974) Measures of pattern complexity for choroplethic maps. *The American Cartographer* 1(2): 159–169

Monmonier M S (1991) *How to Lie with Maps*, University of Chicago Press, Chicago Illinois

Muehrcke P C (1972) *Thematic Cartography*. Commission on College Geography, Resource Paper 19, Association of American Geographers, Washington DC

Muehrcke P C (1973) The influence of spatial autocorrelation and cross correlation on visual map comparison. *Proceedings of the American Congress on Mapping and Surveying*, pp. 315–325

Muehrcke P C (1978) *Map Use: Reading, Analysis and Interpretation*, J P Publications, Madison Wisconsin

Muller J-C (1975) Associations in choropleth map comparison. *Annals, Association of American Geographers* 65(3): 403–415

Muller J-C (1976) Number of classes and choropleth pattern characteristics. *The American Cartographer* 3(2): 169–175

Muller J-C (1992) Visual association in choropleth mapping. *Proceedings of the Association of American Cartographers* 8: 160–164

Muller J-C (1992) Generalization of spatial databases. In: Maguire D J, Goodchild M F, Rhind D W (eds) *Geographical Information Systems: Principles and Applications*, pp. 457–475

Oden N L (1984) Assessing the significance of a spatial correlogram. *Geographical Analysis* 16(1): 1–16

Olson J M (1974) Autocorrelation and visual map complexity. *Annals, Association of American Geographers* 65(2): 189–204

Openshaw S (1992) Developing appropriate spatial analytical methods for GIS. In: Maguire D J, Goodchild M F, Rhind D W (eds) *Geographical Information Systems: Principles and Applications*, pp. 389–402

Peterson M P (1985) Evaluating a map's Image. *The American Cartographer* 12(1): 41–55

Robertson P K (1988a) Choosing data representations for the effective visualization of spatial data. *Proceedings, Third International Symposium on Spatial Data Handling*, Sydney, Australia, pp. 243–252

Tobler W R (1973) Choropleth maps without class intervals? *Geographical Analysis* 5: 262–265

Weibel R, Buttenfield B P (1992) Improvement of GIS graphics for analysis and decision-making. *International Journal of Geographical Information Systems* 6(3): 223–245

13 TIME AS A CARTOGRAPHIC VARIABLE

A. MacEachren

... INTRODUCTION

As computer systems make dynamic interaction with maps more practical and map animations less cumbersome to create, the cartographic perspective on time must change. Time has typically been treated as an attribute to be mapped, with aspects such as length, date of occurrence or change in location over time being depicted much as we might show aspects of vegetation, atmospheric pressure or environmental risk (e.g. Vasiliev, 1991). The representation of time as an attribute does not require time, something that Minard made clear more than a century ago with his now famous flow map of Napoleon's march into and out of Russia. In a dynamic cartographic world, although we can now represent time with time, we may achieve only modest gains by so doing. Conceptualising time only as an attribute limits the potential of dynamic map displays. Time can also be conceived of as a cartographic variable to be manipulated, much as we manipulate size, colour hue and space itself. It is this addition of a powerful variable to our limited cartographic tool kit (Bertin's seven graphic variables) that is likely to make the most substantial impact on maps as a visualization tool in GIS.

Whether static or dynamic, all cartographic representation depends on a limited graphic language which derives from a set of graphic primitives first explored by Bertin (1981). For non-animated maps, Bertin identified seven primitives: size, value, texture, colour, orientation, shape and location in space. Although other authors have added to or modified this list slightly, most cartographers accept the contention that there is a limited set of fundamental graphic variables, illustrated in Plate 13, from which all map representations are built. Bertin specifically excluded movement and time from the system he developed because he felt that a completely different cartographic environment results when these factors are included – a shift from 'graphic system' to 'film'. As demonstrated by DiBiase et al. (1991a), however, Bertin's system of graphic variables seems to hold up in a dynamic environment and this dynamic environment further expands Bertin's static set.

... DYNAMIC VARIABLES

Our research team at Penn State University initially identified three fundamental dynamic variables: duration, rate of change and order (DiBiase et al., 1991b). Here I will add a fourth to that list: phase. As with Bertin's static variables, for which combinations of shape, arrangement, orientation and so on can result in various spatial patterns, combinations of these fundamental variables can result in temporal patterns. For present purposes, however, we will focus on the basic temporal primitives.

DURATION

For static maps, Szego (1987) considered a variety of issues related to how space and time interact and can be conceptualised cartographically. He labelled an instant in time a 'situation'. Traditional static paper maps depict situations. In mapping single situations, duration is independent of the processes of cartographic display. Instead, it is controlled only by how long a viewer wants to look at the map. We can animate an otherwise static map by explicitly controlling duration. When we do so, each situation becomes a frame of the animation (Figure 13.1). A sequence of frames with no change from frame to frame will

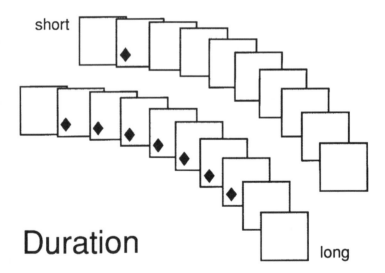

Figure 13.1 An illustration of short and long duration events visible for a few or many frames of an animated sequence

be termed a scene. Szego refers to a coherent sequence of situations (animation frames) as an 'event'. Because duration is a quantity which can be precisely controlled with most animation software, duration of a scene or of a frame can be used to depict ordinal or quantitative data. If animation is applied to a static situation, short duration scenes within events might correspond to insignificant features and long duration scenes to significant features. Applied to dynamic processes, on the other hand, short duration scenes throughout an event imply smooth movement while long duration scenes suggest abrupt movement. The duration of individual frames of an event, or number of frames per unit time, determines the animation's temporal texture. This is referred to as 'pace' in some commercial animation software packages.

RATE OF CHANGE

For events, defined as two or more related situations depicted as animation frames, we obtain a second dynamic variable, the rate of change. From one frame to the next, any of the static visual variables can change, as can the dynamic variable of duration (Figure 13.2). Change in the location variable results in apparent motion and change in any other static variable draws attention to a location and, if change continues throughout an event, can suggest that an attribute at a location is changing. The rate of change can be constant or variable.

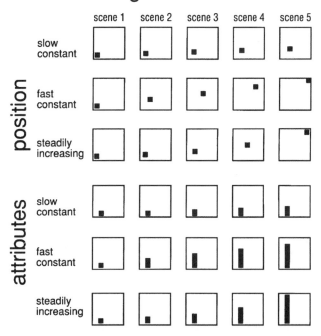

Figure 13.2 The impact of different rates of change in position and in attributes for a five scene animated sequence.

Change in duration of frames in an event can suggest increasing or decreasing importance, or can imply accelerating or decelerating location or attribute change. When attribute or location change is depicted between frames, relative duration of frames can interact with the magnitude of this change. For example, many constant but short duration frames per unit time, with a constantly decreasing change in position of a map symbol, will give the impression of a smooth slowing movement. Frames of geometrically decreasing duration with constant changes in position, on the other hand, will produce an impression of slow abrupt movement gradually giving way to accelerated, explosive, continuous movement.

ORDER

Time is inherently ordered. Matching animation frame order with temporal order of the phenomenon depicted is the most obvious way that order can be used as a dynamic variable. With dynamic maps, however, we can use time order to represent, in a symbolic way, any other order. This is just the same as using an ordered set of grey tones or circle sizes to depict ordered income or population categories or as area is ordered in a cartogram to represent population or some other relevant feature (Figure 13.3). An example of temporal order used to represent something other than time can be found in Slocum et al. (1990). They use presentation order of categories on choropleth maps to emphasise the spatial distribution of each category.

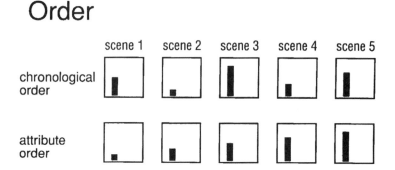

Figure 13.3 Temporal order of animation scenes can be matched to chronological order or to any other order of interest

PHASE

Although my colleagues and I identified three fundamental dynamic variables (DiBiase et al., 1992), a fourth was implied by Szego (1987) in his account of maps that depict space–time phenomena. Szego describes the concept of phase as 'a rhythmic repetition of certain events'. For him, phase is something to be mapped rather than a variable to be used in mapping, but when maps become dynamic, this situation changes. Map animation tools allow the length of time between repetitions (of an order, of a spatial pattern, etc.) to be controlled.

One could argue that phase is a composite of duration and rate of change or order, but, like texture as a static graphic variable, it seems to produce a basic perceptual response that justifies its identification as a distinct dynamic variable. A particularly important application of phase to static animation is its use in colour cycling, one of Gersmehl's (1990) nine metaphors of animation. Colour cycling is usually applied to linear features to suggest movement through those features (Figure 13.4).

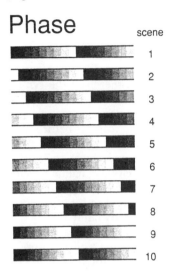

Figure 13.4 Phase applied to colour cycling within a line symbol

...APPLICATION OF DYNAMIC VARIABLES: ANIMATING STATIC MAPS

Conceptually, the lowest level of map animation involves sequencing of spatially and temporally independent scenes to create what Gersmehl (1990) termed a 'slideshow'. In this, the animation presents a series of dissimilar still images such as maps and other information such as graphs or text. The presentation duration of each image can be controlled to indicate relative emphasis and we can take advantage of order to establish logical links between adjacently presented scenes. The prototypical slideshow, however, does not usually use the dynamic variables in this symbolic way and the map base, if there is one, is likely to change abruptly from one slideshow image to the next.

If a sequence of images is spatially dependent, then three categories of application can be identified for the dynamic variables: (1) depiction of existence at particular locations, (2) depiction of attributes at locations, and (3) depiction of change in existence, attributes or location. As will be described below, animating static maps involves a spatially linked sequence of scenes that are temporally independent, while animation of dynamic maps involves a spatially linked sequence of scenes that are temporally dependent.

The concept of animating static maps is interpreted here as the use of animation to depict things other than dynamic processes; the use of time to represent features other than time. Both change in position of the observer and change in position of an object in attribute space seem to fit in this category. In addition, animation can be used in a more traditional symbolic way, on otherwise static maps, to highlight features that do not change. To describe how animation can be applied to static maps, we must begin by considering the three application areas of dynamic variables cited above.

EXISTENCE

'From neon signs in Las Vegas to the blue light atop an ambulance, flickering lights are used in our society to gain attention.' (Travis, 1990) The advantage provided by a dynamic symbol on an otherwise static map is that human vision is particularly sensitive to change, therefore the symbol is noticed and figure-ground is enhanced. We are most sensitive to flicker at about 30 degrees in the periphery of central vision, so that flashing symbols allow us to notice the presence or existence of important map features before we focus directly on them.

Just as any non-locational static variable when applied to a particular location represents the existence of some object or feature at that location (e.g. a star shape representing a capital city), dynamic variables can have this simple symbolic function. Switching between constant duration scenes in which a static graphic variable applied to a symbol or set of symbols is varied from scene to scene is the most basic case of using dynamic variables to highlight existence. In this case, there is no spatial or temporal change from scene to scene. Such a procedure has been demonstrated for point features that depict earthquake locations (DiBiase et al., 1992) and for area features used to highlight geographic homogeneity (or lack of it) in female success at attaining elected office (Monmonier, 1992). In a typical GIS context, individual data layers on composite maps might be blinked on and off to allow a viewer to visually isolate them without removing the other data layers from view.

ATTRIBUTES

Dynamic variables can also be used to symbolise an attribute of a feature being depicted. An innovative use of duration to depict an attribute is Fisher's use of duration to indicate uncertainty in Chapter 19. In one of his applications, duration is used to represent the likelihood that the cells on a gridded soils map have a specific soil type. The more likely that the soil in a grid cell is in a particular category, the longer that cell appears in the hue associated with that type. The resulting map appears quite stable in regions of soil classification certainty and appears chaotic where certainty is low. As he demonstrates, duration of sounds as well as graphic variables can be used to represent attributes at locations.

Colour cycling as noted above is an effective use of phase (of hue and/or value ordering) both to highlight and depict the attribute of direction. The best known use of this technique is on television weather programmes to portray location and direction of the jet stream. Colour cycling has an advantage over moving symbols on a complex map because, as symbols are static, with only an illusion of movement, no attention need be given by the animator to the potential collision of symbols. Like actual movement, colour cycling is a particularly strong mimetic symbol because human vision tricks us with the sensation that we are perceiving motion even though the symbol location remains static. We intuitively associate this apparent symbol movement with physical movement. An extended symbolic application of colour cycling, beyond that used on typical television weather maps, would be to vary the rate of colour cycling as a direct indication of flow magnitude. Tobler's (1981) fiscal flow maps, for example, can be effectively animated in this way (Figure 13.5). On such a map the Gestalt principle of 'common fate' predicts that viewers should be able to group common apparent movements in speed and direction and separate regional patterns if they exist.

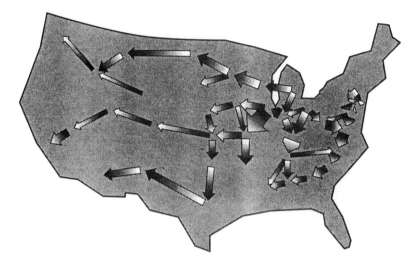

Figure 13.5 One application of phase is the production of colour cycling

A common feature of the dynamic symbolisation of feature attributes cited above is that each could be shown with static symbols on a single static map. None depict change in either space or time, although colour cycling implies it. As with the use of viewing time to depict existence, the dynamic symbols have apparent figure-ground and attentional advantages. They can be used to give particular emphasis to limited aspects of a map and, perhaps more importantly, to change that emphasis over time. For exploratory visualization applications, it is this potential dynamically to alter emphasis that makes animated static maps an important development in visualization for GIS.

CHANGE

Change depicted with dynamic maps can be of three types involving change in the position of the observer in relation to geographic space, shown with viewpoint change, change in position and/or attributes of an object within geographic space, shown with time series and change in position of an object within attribute space, shown with re-expression (DiBiase et al., 1992). Time series are obviously applicable to dynamic spatio-temporal processes and are perhaps the first application thought of when map animation is considered. With time series maps, the sequence of scenes is temporally dependent. (For example the depiction of change in AIDS incidence per county over time requires that scene order be matched to chronological order.) When using dynamic variables to depict change in observer position, on the other hand, the sequence is temporally independent. No particular ordering of scenes is required because what is viewed is not changing over time. Viewpoint change is, however, spatially dependent. Change in location or scale from scene to scene will be perceived as observer movement so it must occur in a logical, spatially sequential, order. Change in attribute space is both temporally and spatially independent. The sequence of scenes is matched neither to chronological order nor position order, but to the order of some attribute at places for a particular time. This use of time to represent orders other than chronological is comparable to the use of space to represent variables other than space advocated by Dorling in Chapter 11. As with use of animation to depict existence or attributes at locations, depiction of change in viewpoint and change in attribute space can be considered an application of animation to temporally static maps.

Viewpoint change can involve zooming and panning to change apparent distance from, and position over, a map. These techniques are usually used to examine a two- or three-dimensional static time slice. One successful example of viewpoint change in a non-geographic context is its use in medical imaging as a diagnostic tool. Closer to a GIS context, user controlled changes in viewpoint have been the fundamental basis of flight simulators. In an environment resembling a flight simulator, Bishop (Chapter 8) uses changes in viewpoint to allow a viewer to traverse a realistic model of the natural environment and see the impact of insect infestation throughout a forested region. In relation to global environmental problems, standard flat map projections provide an interrupted view with false edges. Viewpoint change in relation to a model globe can overcome this limitation by simulating the act of spinning a physical globe (Figure 13.6). Dorling (1992) found changes in viewpoint, which he termed 'animating space', to be quite successful as a way to explore spatially complex two-dimensional maps. Although the use of three-dimensional panning and zooming has proved to be particularly useful in medical visualization applications, flight simulators and Bishop's landscape visualization applications, Dorling (1992) suggests that it may be less useful than two-dimensional changes in viewpoint as a tool for visualizing abstract spatial concepts. He also argues that for visualization of enumeration unit data of the kind typical in urban and regional GIS, the zooming capability of viewpoint manipulation may be even more critical than the ability to pan in complex paths over the surface depicted.

Time series involve change over time in position or attributes of an object rather than in position of the viewer. They can be depicted by using viewing time directly to emphasise temporal order, that is, by change in temporal space. This use of time does not apply to animating static maps, but is described briefly here for completeness of the typology and is considered in more detail below. Time can be treated as a vector with viewing time matched to the duration between discrete time samples. It is this use of viewing time that has been most common in past animations (Figure 13.7). Frame order might be matched

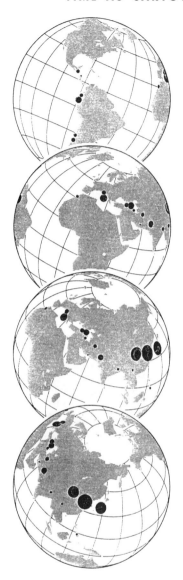

Figure 13.6 A series of animation frames representing the dynamic spinning of a virtual globe

to order of time aggregates such as an hour, a week or a month to produce what Moellering (1976) termed 'collapsed real time'. Gould et al. (1990) used this mapping of aggregate chronological time onto animation time to produce animated map movies of the incidence of AIDS in the western United States from 1981 through 1988. Kabel and Heidl (1992) have extended this analysis to the entire US from 1980 to 1991, with predictions through 1995. Time can also be treated as cyclic, with aggregate values depicted for positions in the cycle. An example relevant to GIS applications in emergency management

123

180 million years B.P.

present

Figure 13.7 Selected scenes from an animated map depicting continental drift

would be the depiction of traffic volume along a road network for composite days. Such a depiction would indicate the typical ebb and flow of traffic and could be used to route emergency vehicles so that peak traffic corridors at a particular time of day were avoided.

Re-expression includes at least three types of transformation relevant to time as a cartographic variable. These are the selection of a subset from time, space or both; reordering based on an attribute other than time; and emphasis based on non-linear mapping of a variable onto time (DiBiase et al., 1992). While temporal brushes (selecting a time subset) can be useful in exploring spatio-temporal processes, other forms of re-expression are applicable to non-temporal analysis and, therefore, represent additional cases of animating static maps. As noted above, order can be applied to depiction of any ordered variable. An interesting reordering example is Monmonier's (1992) combination of order and duration simultaneously to 'sweep-by-value' across a map and bar graph. In this application, a choropleth map of the US showing percentage of female elected local officials appears on a computer screen together with a rank ordered bar graph of the percentages by state. The animation consists of frames having equal duration and constant percentage change from frame to frame. The viewer is presented with an evolving event in which bars on the graph and corresponding states are filled with a 'signature' hue. In parts of the distribution for which percentage of female elected officials is similar, the presentation depicts a smooth filling in of the map and graph. For parts of the distribution in which numerically adjacent states have substantially different percentages, however, the map's evolution appears to hesitate, then jumps quickly ahead in sudden disconnected steps. In a similar application, our Penn State research group has combined reordering and non-linear mappings onto time to help global climate researchers explore issues of global climate model uncertainty about temperature in Mexico (DiBiase et al., 1992). By sequencing temperature predictions by order of their uncertainty and increasing duration of scene presentation in proportion to that uncertainty, researchers viewing the animation uncovered an unexpected pattern of planting season uncertainty in global climate model predictions (Figure 13.8).

Reexpression

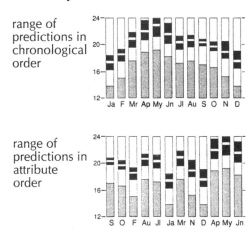

Figure 13.8 Graphing actual and predicted temperature using chronological order emphasises the expected increase of temperature during the spring and summer months and the decrease in fall through winter

... APPLICATION OF DYNAMIC VARIABLES: DEPICTING DYNAMIC PROCESSES WITH TIME

As will be exemplified in more detail in the next chapter, a substantial portion of geographic research focuses on dynamic processes that vary over time as well as space. For visualization tools to be effective, therefore, they must depict the dynamic elements of the processes studied. Restricted to a static product, past cartographers have developed many clever representations of dynamic processes as snapshots out of time such as Smith's Johnstown flood map reproduced in MacEachren et al. (1992). The introduction of time as a cartographic variable offers a more direct way to depict temporal change. Where temporal patterns are expected, or simply possible, depicting time with time in dynamic animation can uncover patterns that remained hidden in sets of static maps. A number of examples from global model outputs of such patterns are shown in the examples cited in Brodlie et al. (1992).

In 1959 Norman J W Thrower reviewed the use of animated maps by the educational film industry and presented cartographers with the challenge of developing maps for this new dynamic environment. Because pre-computer map animation involved manual construction of dozens of individual map frames per second of animation, few map movies were actually produced.

Over the three decades since Thrower directed our attention to the potential of animation, most efforts have focused on producing map movies that were conceptually similar to those Thrower described, that is they were depictions of time as an attribute using a flipbook procedure of sequenced frames. As early as 1970, computers were used to create the individual frames to be photographed (Tobler, 1970). Within a decade, Moellering (1978) developed a system in which simple map image frames could be derived computationally and displayed in real time on a computer graphics monitor. Moellering used available tools to demonstrate possibilities for both depicting a dynamic process of diffusion and for conducting a dynamic analysis of 3D time slices, the application of viewpoint change to terrain analysis. With the latter, for even moderately complex terrain surfaces the system required batch generation of image sequences that could then be played back in real time, or recorded on film or videotape for subsequent playback in a technique termed real-time-later (DiBiase et al., 1991b). Recent microcomputer map animations have also relied heavily upon real-time-later sequencing of pre-created two-dimensional maps to represent dynamic processes (e.g. Gersmehl, 1990; Gould et al., 1990).

Microcomputer software tools such as Hypercard now allow flexible access to pre-created (real-time-later) 2D and 2.5D image sequences supplemented by static maps, other graphics and explanatory text (Marshall et al., 1990). Monmonier (1989a) has introduced the idea of graphic scripts, which are temporally sequenced multi-window graphic presentations, as a way to facilitate access to these maps, graphics, and text for exploring a spatial dataset. In investigating the potential of graphic scripts his ultimate goal is to automate the generation of 'a meaningful sequence of views of a multivariate spatial-temporal dataset'. In addition to this real-time-later interactivity, technology at the microcomputer level has now progressed to the point where 2D and 2.5D images can be calculated and displayed on the fly, thus bringing the possibility of interactive manipulation of spatio-temporal displays to all GIS users. Dorling (1992), for example, has developed a visualization environment that allows analysts interactively to control the pace and direction of two-dimensional dynamic depictions of election results for the UK during the period of 1959–92. When three-dimensional depictions using techniques such as ray tracing were produced, however, Dorling found interactive manipulation to be impossible because a few seconds of motion could take many hours to render.

As technology has progressed from real-time-later playback through real-time-later interactivity to real-time interactivity, interest in linking map animation to process models has increased. Corresponding to the three modes of map animation cited, Marshall et al. (1990) identified three categories of visualization techniques involving the application of animation to simulation models called post-processing, tracking, and steering. Post-processing involves the use of visualization tools to explore the results of a model run after the run has been completed. With post-processing, interactive exploration of data generated at various stages of the model and animated time series of the model's stages are possible. Based upon the success of our post-processing cartographic visualization efforts at Penn State thus far, my colleagues and I have argued that post-processing is the area of visualization where cartographers have the most to offer (DiBiase et al., 1992). Our re-expression of global climate model data cited above provides a clear example of the extent to which visualization tools allow post-processing to go well beyond simply displaying model results.

Tracking involves real-time display while a model is running. Although some interactive control over what is being displayed is possible, tracking allows no manipulation of model parameters. It also does not allow the flexibility of analysis provided by post-processing because analytical procedures can take no longer than it takes the model to run. The main goal in tracking is to notice features that indicate the model is not proceeding as desired. For complex models that take considerable time to run, tracking allows the analyst to abort a run and begin again with new parameter settings without wasting time for the model to complete its run. A variation on tracking is the playback of real-time-later sequences of model runs that require substantial computation time. This is the equivalent of time-lapse photography and allows a very slow process model to be observed in human time.

Steering, as the term implies, allows an analyst to interact, not only with the data produced by the model, but with the model itself. Most importantly, steering allows change in parameters, on the fly, in response to visual feedback about the model's progress. Marshall et al. (1990) present results of their research on visual steering of a turbulence model for Lake Erie. Even with a relatively small database and their Cray Y-MP8/864 and Stellar GS1000 supercomputer environment processing speed/resolution trade-offs were encountered. The model could be run for a 2km grid and visually steered in real time using two-dimensional maps and graphics. To obtain near-real time interaction with the model and a three-dimensional visual display, however, resolution had to be reduced to a 5km grid.

... DISCUSSION

Geographers are seldom interested in a static view of the world, but until recently, that was what cartographic visualization tools provided. Successful static depictions of the world's dynamic nature have often received high praise because of the ingenuity required to meld time and space on a static page. Animation has the potential to increase our success at relating time with space because it provides a new cartographic variable that can depict dynamic processes in an intuitive way and increase the possibilities of exploratory data analysis. Three characteristics of features that can be depicted through manipulation of viewing time have been identified: feature existence, attributes of features and changes in feature existence or attributes. Change can be depicted with viewpoint change, time series and re-expression. Although a sequence of views captured on videotape and shown to an audience works well for viewpoint change and time series, for re-expression the impact will probably be best when visualization is interactive, controlled by the viewer rather than by the cartographer. For exploration of three-dimensional solid models such as geological structures, viewpoint change has proven to be particularly useful. With process model tracking and steering, time series is probably the most applicable use of viewing time. It is for post-processing of model results, where time is available to ponder relationships at leisure, that dynamic re-expression is likely to make the greatest contribution.

Maps designed to support GIS-based visualization have very different goals from past maps. Rather than striving for the most accurate map or the single most effective display, visualization tools used in GIS must be flexible and allow exploration of data from multiple perspectives. When a scientist uses a GIS to explore topics such as forest fires, regional unemployment or traffic accidents, there is no predetermined message, only patterns, anomalies and relationships to discover. Identification of appropriate questions to pose will take precedence over information communication.

Particularly in cases where unexpected patterns and relationships may be as important, or more important, than seeing what was expected, time must be used in intuitively logical ways. In order to realise the potential of time as a cartographic variable, its relationship to non-temporal cartographic variables needs to be understood. It should prove useful to consider this new variable in relation to Bertin's (1981) semiology of graphics. Although Bertin argues that his system for linking graphic variables with referents does not apply to dynamic maps, DiBiase et al. (1991a) have provided evidence to the contrary. Once we allow time to become a variable, we also create the potential for audio variables which require a finite time to be heard. There is now a need to delineate a set of audio variables for use in interactive exploratory analysis. Krygier (1991) and Fisher (Chapter 19) have begun to address this task.

Adding time to the cartographic toolkit has several specific implications for the role of maps in GIS. As an extension of GIS analysis, viewpoint change can be useful as a presentation tool to help the public understand the visual impact of a proposed highway decision or of a disease infestation in a forest. Time series are, of course, applicable to the direct depiction of spatio-temporal processes. Using viewing time to represent real time makes intuitive sense and should be easy for non-expert viewers to grasp in situations where there is only one chance to see the resultant maps as in animated map movies. Spatio-temporal processes can be uncovered when a GIS incorporates time as well as space in its data structure. For exploratory uses of GIS, addition of time to our toolkit of display variables will substantially extend our visual analysis capabilities. The ability to map non-temporal attributes onto duration, rate of change, order and phase makes multivariate visual analysis much more practical than when that analysis is limited to the seven to ten static graphic variables to which we are usually restricted.

... ACKNOWLEDGEMENT

The basic conceptual structure presented here was developed as part of a seminar given by the author and David DiBiase (with substantial input from John

Krygier, Catherine Reeves and Alan Brenner who were enrolled in the seminar). A pair of video essays was produced by participants to illustrate this conceptual structure (DiBiase et al., 1991a). Other applications of the conceptual structure are given in DiBiase et al. (1991b) and DiBiase et al. (1992). The author wishes to acknowledge the substantial role played by his colleagues in the evolution of the ideas presented here, but also to remind readers that any flaws in logic must be attributed to himself.

··· REFERENCES

Bertin J (1981) *Graphics and Graphic Information Processing*, Walter de Gruyter, Berlin

Brodlie K W, Carpenter L A, Earnshaw R A, Gallop J R, Hubbold R J, Mumford A M, Osland C O, Quarendon P (eds) (1992) *Scientific Visualization: Techniques and Applications*, Springer-Verlag, London

DiBiase D (1990) Visualization in the earth sciences. *Earth and Mineral Sciences*, Bulletin of the College of Earth and Mineral Sciences, The Pennsylvania State University 59(2): 13–18

DiBiase D, Krygier J, Reeves C, MacEachren A, Brenner A (1991a) *An elementary approach to cartographic animation*. Video presented at the Annual Meeting of the Association of American Geographers, April 13–17

DiBiase D, MacEachren A, Krygier J, Reeves C, Brenner A (1991b) Animated cartographic visualization in earth system science. *Proceedings of the 15th International Cartographic Association Conference,* Bournemouth UK, pp. 223–232

DiBiase D, MacEachren A, Krygier J, Reeves C (1992) Animation and the role of map design in scientific visualization. *Cartography and Geographical Information Systems* 19(4): 201–214

Dorling D (1992) Stretching space and splicing time: from cartographic animation to interactive visualization. *Cartography and Geographical Information Systems* 19(4): 215–227

Gersmehl P (1990) Choosing tools: nine metaphors for map animation. *Cartographic Perspectives* 5: 3–17

Gould P, DiBiase D, Kabel J (1990) Le SIDA: la carte animee comme rhetorique cartographique appliquee. *Mappe Monde* 90(1): 21–26

Kabel J, Heidl R (1992) *AIDS in the United States, Past, Present, and Future, 1980–1995,* unpublished video

Krygier J (1991) Sound variables, sound maps, and cartographic visualization. Draft manuscript available via electronic mail from JBK5@PSUVM.PSU.EDU

MacEachren A M (1992) Visualizing uncertain information. *Cartographic Perspectives* 13: 10–18

MacEachren A M, Bottenfield B, Campbell J, DiBiase D, Monmonier M (1992) Visualization. In: Abler R, Marcus M and Olson J (eds) *Geography's Inner Worlds*, Rutgers University Press, New Brunswick, N J, pp. 99–137

MacDougall E Bruce (1992) Exploratory analysis, dynamic statistical visualization, and geographic information systems. *Cartography and Geographical Information Systems* 19(4): 237–246

Marshall R, Kempf J, Dyer S, Yen Chieh-Cheng (1990) Visualization methods and simulation steering for a 3D turbulence model of Lake Erie. *Computer Graphics* 24(2): 89–97

Moellering H (1976) The potential uses of a computer animated film in the analysis of geographical patterns of traffic crashes. *Accident Analysis & Prevention* 8: 215–227

Moellering H (1978) *A demonstration of the real time display of three-dimensional cartographic objects*. Computer-generated videotape, Department of Geography, Ohio State University, Columbus Ohio

Monmonier M (1989a) Geographic brushing: enhancing exploratory analysis of the scatterplot matrix. *Geographical Analysis* 21: 81–84

Monmonier M (1992) Authoring graphic scripts: experiences and principles. *Cartography and Geographical Information Systems* 19(4): 247–260

Slocum T A, Robertson S H, Egbert S L (1990) Traditional versus sequenced choropleth maps: an experimental investigation. *Cartographica* 27(1): 67–88

Szego J (1987) *Human Cartography: Mapping the World of Man*, Stockholm: Swedish Council for Building Research

Thrower N (1959) Animated cartography. *The Professional Geographer* 11(6): 9–12

Tobler W R (1970) A computer movie simulating urban growth in the Detroit region. *Economic Geography* 46(2): 234–240

Tobler W R (1981) Depicting federal fiscal transfers. *Professional Geographer* 33(4): 419–422.

Travis D S (1990) Applying visual psychophysics to user interface design. *Behavior & Information Technology* 9(5): 425–438

Vasiliev I (1991) The dimensionality of time in cartographic thought. Paper presented at the Annual Meeting of the Association of American Geographers, Miami, Florida

14 Some ideas about the use of map animation as a spatial analysis tool

S. Openshaw, D. Waugh & A. Cross

... Introduction

Recent developments in accessible multimedia technology have allowed spatial and statistical analysis methods to be employed as a means of improving visualization in GIS and particularly in the animation of time-rich map databases. The immediate problems are twofold, involving, first, the motivations for wanting to approach spatial analysis in this way and, second, how to do it. This chapter suggests and illustrates some ways of overcoming them which may form a useful basis for future research.

... Background

The information technology revolution has meant that more and more data associated with the management of society have been computerised. A particularly relevant feature has been the increasing use of computers not only to record phenomena but also to assign to them an accurate time and date. The last two decades have also seen the proliferation of GIS, which has meant that information is increasingly more accurate in its geographic referencing. An example of such a modern database is that held by Northumbria Police, who record not just the type of incident but its location to within a spatial precision of 100m and the time of reporting to a precision of one minute for every incident they respond to. However, to date most geographical information has been regarded as static and until recently few space–time data series existed. Now a major challenge facing GIS is how best to tackle the geographic data explosion while at the same time incorporating the valuable temporal element which logically accompanies a majority of computerised databases.

... Visualization and animation in spatial analysis

There are now rich datasets, with accurate time and space referencing, and high speed computers capable of processing them. From a spatial analysis perspective the answer is clear. We need to devise an exploratory technology to assist in searching for patterns and processes in GIS databases which would not have been immediately obvious if mapped in conventional ways. Animation in particular, and visualization in general, provide useful creative ways of handling vague questions such as:

What is going on?
Is there anything interesting?
What is happening where and when?
Are there any discernible patterns?

With the exceptions of disease epidemiology and regional forecasting, space–time analysis and modelling procedures are not generally well developed at present. However, neither of these modelling techniques is useful in the analysis of more general space–time databases, and methods have yet to be developed for coping with the tremendously detailed data in which time is measured in units of minutes and spatial coordinates are measured to 100m or less. After all, until recently data with these characteristics have not existed! This is complicated further by the fact that data often occurs in context, and is related to phenomena for which there is little or no previous research to guide the analysis, no process understanding and only limited theory to aid model specification. It is not known what statistical models, if any, may be applicable. Commonly, there are only the vaguest of notions about what to expect and limited understanding about the nature of the patterns and processes that may exist.

The standard geographic way of coping with arbitrary space–time datasets has been to aggregate the data both geographically (into a small number of large and easily manageable areas) and temporally (into a small number of time periods). Having done this, choropleth maps have been drawn for differing time periods and comparisons made between them. This reductionist strategy converts an otherwise extremely rich and detailed space–time data series into a relatively poor quality space–time dataset which only serves to destroy most of the information involved.

... EXPLORING THE USE OF VISUALIZATION

It would seem that many chance discoveries of process have been made, often accidentally, by an expert mind observing what is going on around it. Good ideas, theories and hypotheses have been generated by abstract visualization of pattern and process. In the absence of concrete theories and models to cope with these new space–time datasets, it is necessary to identify some alternative ways by which the expert can look at the data, in the hope that theories and models may spring from their observations. The remainder of this chapter describes one possible way of doing this using simple movie-making. It includes a brief description of an algorithm developed by Dorling and Openshaw (1991) and some preliminary results from the space–time analysis of a standard police incident log database. Many of the concepts outlined are similar to, and develop from, those of the visualization group at Pennsylvania State University outlined in the previous chapter.

... COMPUTER MOVIES

The idea of computer animation of time driven map sequences dates back to the mid-1960s. Tobler (1970) provides an excellent early illustration of the potential that exists for movies in the visualization of time-dependent geographic data. He summarises very succinctly the objective of what is now known as visualization: 'Because a process appears complicated is no reason to assume that it is the result of complicated rules.' (p. 234) In particular, time-dependent spatial patterns may appear extremely complicated but when viewed on film but with an appropriate level of abstraction their essential simplicity may well emerge for all to see. In Tobler's early work he had to start by inventing some data which involved building a time explicit model of population for an urban region and then copying the resulting sequence of static maps for each time period to a small number of frames of cine film. The resulting playback at normal viewing speed neatly summarised urban population change for the period of interest. The weakness of Tobler's original movie was in the model employed to generate the time-dependent data. It was a movie not of reality but of the workings of a model.

Even as early as the 1970s some fine resolution space–time data series existed and the film by Moellering (1973a; 1973b) of traffic accidents in Washtenaw County from 1968 to 1970 demonstrates its potential as a spatial analysis tool. Moellering's movie is based on a large number of traffic accident maps and provides a visual impression of the spatial patterns and intensity of accidents. By playing around with the design of the movie it would have been possible to search for patterns that might better inform transportation engineers. For example, repeated accidents at the same location or major changes in location and intensity with time might be detected without making any prior assumptions.

Tobler and Moellering's procedures can be seen to be driven by a basic movie-making algorithm which involves the following steps:

1 Obtain a space–time data series, ideally with a few thousand time intervals such as minutes in a day.
2 Select an appropriate time increment, dividing the series into several hundred time periods, for example, five-minute intervals.
3 Produce a map for each time period.
4 Record the map on one (or more) frames of film or video.
5 Repeat steps 3 and 4 until all time periods are processed.
6 View the results.

The principal problem with this was the time and effort required to produce a hundred or so maps and record each one onto film or videotape in a controlled fashion and perhaps this is why movie-making never seemed to realise its full potential as a tool for spatial analysis. However, with the arrival of cheap and powerful computer hardware, the procedure can be both fully automated and enormously speeded up. This, coupled with the ability interactively to change various design aspects of the movie almost in real time, creates the prospect of an exciting new method that relies on visualization of animated maps rather than on statistical analysis.

The computer movies discussed in this section have been produced by two separate computer programs. The first is used to produce a highly compressed

133

version of the movie on disk file. The second is used to decompress this file in real time to the computer display. Optionally, the output from the computer can be connected to a video rather than a monitor. It is assumed that the data relate to events at points in time and space such as reported crimes or occurrences of a particular disease.

The first program generates a smooth map display by applying to each time period a three-dimensional kernel density estimation procedure that smooths the data in both space and time (see Chapter 9), spreading the influence of each data point around where it occurred. The further from the actual point in space–time, the less influence of that point on the density value. Next, the program 'colours' the map according to the total height of each pixel based on all the kernels intersecting at that particular pixel before compressing the resulting map and saving it on disk. These two steps are repeated until all time periods have been exhausted.

The second program is used to read these disk files, decompress them and display the movie on screen. It is necessary to store movies in compressed format because a decompressed movie would take up far too much space on the disk (a 256 frame movie, where each frame is 640 × 480 in 16 colours, would require 37.5 Mb) and would also make updating the display far too slow.

It is worth emphasising that the final movie may well reflect key design decisions which have been involved such as the choice of smoothing parameters. It is important to experiment and to animate random time versions of the same data (that is keeping the spatial coordinates the same, but scrambling the time coordinates). This will help to find meaningful patterns, not just ones which are artifacts of the movie-making process.

A movie for leukaemia incidence has been produced using 680 cases of childhood leukaemia occurring over the last twenty years in the North of England. For these data, density estimation was accomplished using a spherical kernel with a 25 km radius in space and a 12 month 'radius' in time. The movie showed that, as would be expected from their population densities, the worst areas for leukaemia were Manchester and Newcastle. However, during viewing a striking increase in cases was seen in Middlesbrough which lasted for just a few years. Amalgamation of all the years experiencing low occurrences with those experiencing high ones would almost certainly have destroyed this anomaly. Similarly, it was also seen that over the last ten years there was an oscillation between rises and falls of occurrences in Newcastle and Manchester. This occurred four times and while it may be some kind of artifact of the movie-making process, it could also be something more significant. As a final note, it should be pointed out that, perhaps surprisingly, there were no significant patterns around Sellafield nuclear power station during this time, yet this is a site which is often quoted as the main source for an observed cluster of cases.

Replaying mapped data in this manner therefore allowed the viewer to spot spatial phenomena and/or any systematic changes in spatial patterns over time which may not otherwise have been apparent from any static data, including processes such as diffusion, flows and oscillations through time. However, it must be stressed that any spatial analytical insights are a direct result of the viewer's conscious or subliminal identification of key patterns. The movie procedure is simply a tool to help the user spot patterns, it will NOT spot patterns for itself. It should also be stressed that the analysis of space–time datasets is fraught with all manner of difficult technical problems and the temporal aspect only serves to complicate any formal analytical procedures further. However, this problem may be alleviated slightly by presenting data visually in an animated form. The challenge therefore is to discover how animation should best be used in order to enhance the understanding of data.

... HOW ANIMATION SHOULD BE USED

As explained in the previous chapter, the ordering of maps is a key design feature. Three alternatives are:

- Speed up time, but retain the temporal structure of the data in a search for time-driven spatial patterns.
- Reorder the data by a subfield within time, for example by hours and minutes, ignoring the day, month and year. This should promote the visualization of time-of-day-dependent patterns, but other spatial patterns may well become scrambled by it.
- Order the maps based on a variable other than time. An example would be cancer incidence maps ordered by increasing the amounts of radionuclides released from a reprocessing plant. Again, as the previous chapter illustrates, it is certainly useful to escape from a narrow definition of time as a data ordering device to investigate other variables that are not necessarily functions of time. Whether the resulting map jumble yields any identifiable patterns is an issue that will be application-dependent.

There are a number of enhancement procedures that could be applied to reduce some of the complexity present in the mapping of time driven spatial data. These include:

- A moving window approach. In this the map for a time period depends not just on the data from that period but also on those from the (n−1)th, (n−2)th, (n+1)th (n+2)th, and so on, with the degree of smoothing controlled by the width of the window chosen. There may well be some relationship between the width of the smoothing window and the detection of spatial pattern. Smoothing data in this way is useful in avoiding patterns being suggested where there is in fact no strong pattern. It should be pointed out that this windowing process is already present in the way the data are initially clumped into periods before the movie is produced. The question is really, what, if any, further smoothing would be helpful.
- An alternative windowing procedure is to start at the first time period and put all data points occurring within that interval on the map. This map is then recorded as the first frame of the movie. Next, all the data points occurring within the second time period are added to the map and the result recorded as the second frame. This is repeated for each subsequent time period. When played back the movie will show the point data gradually building up with all of the data shown on the last frame.
- Another procedure allows the luminosity displayed to decay progressively with time until the points or symbols disappear altogether after some fixed time. For point data this can be accomplished using a linear or exponential decay function based on the results of experimentation. For zonally based

135

displays a smoothing function, as in time series analysis, can be applied. In both cases the objective is to provide what is displayed with a shadow or memory of past time periods in the hope that this will add some sense and meaning to the data.

Problems which remain and are subject to further discussion relate to:

- selecting appropriate colours for map display;
- determining the appropriate film speed film in order to maximise perceptual insights;
- investigating different graphic presentations including symbolism and other design aspects which can influence user perceptions of map patterns and associated changes;
- understanding map perception and cognition processes, both from the stance of the animator and the viewers.

Essentially, and as shown in the previous chapter, all these design variables can be adjusted by the animator as a means of refining, restructuring and manipulating maps which they have created and it is argued here that 'the idea is to learn through creating and not only from the creation' (Dorling and Openshaw, 1991, p. 396).

One other possibility involves the sound channel on the video which might be used to contain additional pattern describing information. At its most elemental level, music consists of pitch, duration and dynamics. The view expressed here is that if performed in an appropriate fashion 'map music', or more correctly 'sounds' relating to the spatial analysis being performed, may provide an additional stimuli. The latter could be used to represent some aspects of the analysis which cannot necessarily be seen on the map display, for example varying the pitch according to the statistical significance of the map patterns of using dynamics (or intensity of sound) to represent the number of cases on the map or some other measures of gross map pattern intensity. In Chapter 19, Fisher illustrates the 'sonification' of a single map to convey information on the accuracy of a classification.

Whether or not the viewer could actually tolerate such a powerful integrated audiovisual experience is left for further research. The suggestion is that some kind of audio reinforcement of spatial pattern might be a useful aid for the analyst. Sound is already used to good effect in Geiger counters and sonar and is inherent in virtual reality, so there is no reason why it might not also work here. It is certainly worth experimenting with even if at first sight it may appear a little strange.

...A CASE STUDY: POLICE INCIDENT LOG DATA

In this section, we illustrate these ideas using as an example the production of movies based on incident log data for the whole of 1989 supplied by Northumbria Police. These data consist of incident types, a 100 metre grid reference, and the time that the incident was reported to the nearest minute. The example includes a description of the approaches adopted to analyse these data using animation. It is not an exhaustive list of what has been achieved using animation, but it does give a flavour of the use of movies for investigating important and relatively unexplored databases.

In this first experiment all the incidents for the whole of 1989 for Tyne and Wear were sorted according to the time at which they were reported but with the date completely ignored. Movies were produced for various incident types and for each a frame was produced for every five minute interval during the standard day. The kernels used were spherical with a spatial radius of 1500m and a thirty-minute temporal 'radius'. Although several movies for the various incident types were created, it was clear that, because the data give the time of reporting rather than the actual time of the crime, the view portrayed was one of police workload rather than criminal activity. For example, the movie for incidents of burglary other than from dwellings (ie. business premises) high-lighted the city centre and business areas and the most dramatic increase was observed at about 8 a.m. when break-ins were discovered and died away as the morning progressed. In the case of public disorder offences the movie was far more informative because the reported time is a better indication of the time that the incident was committed. Plates 14 to 16 can only show snapshots from this movie. Plate 14 shows the expected peaks in public disorder incidents which occur when public houses and favourite night spots are closing, while Plate 15 shows the quieter periods during the morning. Finally, Plate 16 shows a time period in the early evening with high incidents of disorder which was initially met with surprise by the police. After some thought, the police decided that this peak represented juvenile disorder, with youths wandering the streets, getting into fights and generally causing a nuisance. This provides a clear exam-ple therefore of how animation can be effectively combined with an expert's knowledge providing greater benefits and insights into the space–time rich databases available.

In another study every public disorder offence within a ten-mile radius of the Newcastle city centre weather station was assigned a temperature. Each incident temperature was found by looking at the time the incident was committed and assigning the respective hourly temperature from a weather database. This was carried out for the whole of 1989 and a map was produced for every 0.1K from 3 to about 25°C. The data were normalised to give rarer temperatures a higher weighting. This weighting was necessary because only three or four hours in the whole of 1989 experienced temperatures in the region of 25°C, whereas there were numerous incidents recorded during a registered temperature of 10°C. Thus one would expect there to be far more incidents occurring at 10°C than at 25°C, but weighting overcomes this problem. The resulting movie shows that, as might be expected, public disorder increases with temperature.

...CONCLUSIONS

At present most spatial analysis is done blind. Maybe the main importance of animation at this stage is to reduce the degree of blindness. It is suggested that visualization of data by animation offers the basis for a new approach to spatial analysis, providing a computerised version of the inductive process of science. Temporal dynamics, long viewed as a confounding factor in spatial analysis because of the problems it causes to statistical analysis, is now viewed as a potentially useful analytical tool.

The whole appeal of computerised visualization is that it provides an artistic, creative, spatial analysis technology that may be extremely useful in a wide range of circumstances and applications. It retains the communication power of the map, but injects into this an element of dynamism. In some applications the insights that are gained may be built into computer models and theories, in others there may be no need for any other form of analysis because visualization is itself sufficient. This is further enhanced by the fact that the necessary basic technology is becoming cheap, simple and widely applicable, so that virtually any space–time dataset can benefit from this approach.

The argument is that the use of video technology and the observation of data in 'all its glory' may afford a mechanism whereby it is far easier to 'see' patterns and processes by watching a movie than it is to model them using mathematics, or by observing similar processes in the real world.

⋯ REFERENCES

Dorling D, Openshaw S (1991) Experiments using computer animation to visualize space–time patterns. In: Klosterman R E (ed.) *Proceedings of Second International Conference on Computers in Urban Planning and Management*, Oxford, pp. 391–406

Moellering H (1973a) The automated mapping of traffic crashes. *Survey and Mapping* 33: 467–477

Moellering H (1973b) Computer animated film available on traffic crashes Washtenaw County Michigan 1968–70. University of Michigan, Ann Arbor, Michigan

Tobler W R (1970) A computer movie simulating urban growth in the Detroit region. *Economic Geography* 46: 234–240

SECTION C
VISUALIZING DATA VALIDITY

15 INTRODUCTION TO VISUALIZING DATA VALIDITY

M. GOODCHILD, B. BUTTENFIELD & J. WOOD

... BACKGROUND

First and foremost, effective visualization of data in GIS requires the use of an appropriate method of representation. In general, the validity of a graphical display may be ensured by application of sound principles of graphical design, or evaluated for cognitive effectiveness, and each of these options is discussed in other sections of this book. In addition, geographical data are generally incomplete, and almost always have some degree of uncertainty. Thus while much of this book is concerned with the issue of valid representation, its converse, the representation of the validity of the data, is of particular significance to spatial data. This forms the focus of this section.

Information on validity affects the reliability of all aspects of analysis, spatial inference and reasoning. It also affects the credibility attached to decisions supported by geographic information systems (GIS). Collection and maintenance of valid spatial data has long been a goal of data producers, but only recently has the incorporation of data validity become a priority. In the context of GIS, the term 'validity' encompasses concepts related to measurable validity, including, but not limited to, measurement by deductive estimates, inferential evidence, or comparison with independent sources. Measurement and evaluation may include indices and processes, and visualization of these two types will be presented in the chapters in this section.

Dictionaries define the term 'valid' using phrases such as 'sound or defensible', 'well-grounded', 'executed with proper formalities (or formalisms)' and 'legally acceptable'. A search for the word 'validate' recovers words like 'ratify', 'confirm' or 'verify', implying that validity can be codified in formalized expressions, which in the course of geographical analysis may be numerical, verbal or graphical. Validity can thus encompass both the accuracy of data and of the procedures applied to those data, and becomes especially important when basing interpretation, inference or decision-making upon GIS data. The validity of data may be ratified by testing or by illustration, or by replication. Unfortunately, GIS operations are rarely replicated and this complicates the issues raised in the chapters of this section.

The term 'data quality' encompasses more than testable elements subsumed here under validity, incorporating aspects of lineage that are often monitored or tracked in database operations. Trackable elements include chronological reporting of data collection and processing, algorithms used to process the data, and tests applied to the evaluation. The need for tools to facilitate description, storage and representation of such information has been recognised by the US National Center for Geographic Information and Analysis (NCGIA), which in June 1991 began a research initiative 'Visualizing the Quality of Spatial Information' (Beard, Buttenfield and Clapham, 1991). Priorities for the initiative include development of frameworks linking spatial data quality components with visualization methods, implementation of prototype displays, and database management to support error tracking and to facilitate graphical depiction of error propagation during GIS operations. This section reports research in all three areas.

... VALIDITY IN GEOGRAPHICAL INFORMATION

By definition, reality is continuous, while the observation of reality is discrete. Technology discretises measurement, as for example in satellite image 'snapshots' taken at regular intervals in comprehensive scanning paths. Perception also occurs in discrete 'chunks', is selective and easily masked or distracted. Digital organisation of data requires that models be fitted to observations and measured phenomena, and, as each of these chapters demonstrates, the model chosen for digital data will to a large extent drive and even constrain the type of visualization that may be appropriate to its subsequent representation. Because it is not possible to represent continuous phenomena completely, they must be approximated, or sampled as subsets. In numerical analysis, this may take the form of discrete approximation formulae, while in modelling, it may occur through processes of sampling, estimation and inference. In mapping, we achieve the same fundamental objectives by processes of abstraction, selection, and generalisation.

The types of validity that may be discovered or measured in geographical data most commonly include error and accuracy. As discussed by Buttenfield and Beard in Chapter 16 these are not identical concepts. Accuracy measures the discrepancy from a modelled or an assumed value, while error measures discrepancy from the true value. One consequence of the discretisation of reality is that true values for geographical phenomena are rarely determinable. It is impossible to collect a complete dataset, so that a 'true' value may be inferred but it is not often determined with full confidence. For this reason, many standards for data quality assessment include indices of accuracy rather than error, and require explicit reporting of the models applied. A common situation in spatial analysis is the measurement of root mean square error (RMSE), which is in fact an accuracy measure of perpendicular distance using a Euclidean model of geometry as reference.

A different aspect of validity relates to replicability, or consistency, of processes and realisations. GIS operations are rarely replicated in practice. The selection of an optimal site or of an optimal route is not usually repeated hundreds of times with slight variations in the values of coefficients or input parameters to determine the sensitivity of the solution to varying inputs. In soil mapping, for example, it is widely accepted that mapped boundary lines contain an element of subjectivity and uncertainty, but rarely have many photoint-

erpretations been made of the same area to evaluate variation between interpreters, or to determine overall 'average' boundary line locations for soil
parcels. Soils mapping proceeds using the educated interpretive skills of a single interpreter, using source data (photos and maps) drawn from multiple
scales at many dates and with varying accuracy. When the 'true' soils boundaries cannot be determined, it is difficult to select a numeric model against
which to compare the interpreted map boundaries, and determination of soils
mapping accuracy remains problematic.

... THE REPRESENTATION OF UNCERTAINTY

Clearly, it is important to convey an appreciation of the uncertainty present in
the data we use in GIS. Moreover, when using visualization as an aid to this
appreciation we are actually visualizing the validity of our data. The measurable aspects of data uncertainty will probably have a spatial component, since
quality will be higher in some areas than others, or higher for certain features.
Graphic representation provides an ideal mechanism for the communication of
this spatial variation of quality.

Although data validity may be measurable, data quality has been defined as a
point in the three-dimensional space of data goodness, application and purpose
(Beard, Buttenfield and Clapham, 1991). A significant consequence of this is
that the data themselves do not contain all of the information required to
locate data quality. The two-way interaction which is a characteristic of ViSC
has enormous potential in allowing the user's knowledge of purpose to be combined with the characteristics of the data.

Despite the many reasons, if not the imperative, to consider data quality
issues, many users do not consider the implications of uncertainty in their use
of GIS. Visualization can act as an effective communication medium, conveying
the effects of error on modelling the real world. This may perform an important role in the education of users in 'data-quality awareness'.

... MODELS OF UNCERTAINTY

As argued earlier, the validity of spatial data derives from many potential
sources and can be subject to a bewildering array of influences. However, visual representations of validity must be controlled by well-defined sets of rules.
One such set of rules frequently used in discussions of uncertainty, particularly
in the context of error or inaccuracy, is provided by statistics, and is the basis
for Chapters 17 and 19 by Goodchild et al. and by Fisher.

In essence, statistics provides a rigorous mathematical framework for
describing and manipulating uncertainty. While the term often conjures up
images of tossing dice and drawing coloured balls from urns, the application of
statistical methods does not imply an assumption that uncertainty in spatial
data similarly derives from such processes. Instead, the assumptions behind statistics are much broader, and relate more to the availability of information than
to any mechanical conception of process. Randomness simply reflects lack of
knowledge, and statistics provides a means of dealing systematically with

incomplete information. For example, suppose we are uncertain about the true position of an object because we cannot determine which datum was used when its position was measured, or which projection was used for the map from which its position was digitized. These possibilities reflect different uncertainty-causing processes. Statistics provides a means of handling uncertainty that is responsive to its form but independent of the process by which it was generated. From this viewpoint, uncertainty may be temporary, since we might later discover the missing information on datum or projection.

From this perspective, the statistical paradigm is potentially useful for all sources of uncertainty and all kinds of validity but its applications are nevertheless limited and we would not want to suggest that it is the only possible framework for visualizing validity. There are circumstances where errors are sufficiently large or unique, and where the generalizations of statistics would have no relevance. The use of statistics also implies that uncertainty is quantifiable, whereas the validity of spatial data is often recorded qualitatively.

Statistics provides a framework for the description and modelling of uncertainty, and thus for its visualization. Uncertainty is described using such concepts as standard error, or root mean square error (RMSE), by measuring the differences between observation and a model, and summarising them as a standard deviation. Commonly used quantitative error descriptors for spatial data include the Circular Map Accuracy Standard (CMAS) (Goodchild, 1991), the Perkal epsilon band (Blakemore, 1984), the misclassification matrix (Chapter 12) or the root mean square error measure used to describe the quality of digital elevation models (US Geological Survey, 1987).

These measures describe the average case, and they can be used as the basis for methods of visualization. For example, a contour map might be accompanied by a map showing the variation of the root mean square error of elevation inherent in the data used to draw it. Many applications of geostatistics (Cressie, 1991) include such displays because Kriging, a standard method of contouring or spatial interpolation, provides estimates of the uncertainties associated with its predictions. Nevertheless, displays of measures of the average uncertainty are of much less value in applications where the focus is on specific errors. In such cases it may be much more valuable to illustrate the range of possibilities, but to do so we must be able to model uncertainty in addition to describing it.

An error model is defined here as a stochastic process capable of simulating the range of possibilities known to exist in spatial data. These possibilities may exist because vital information, such as datum or map projection, is missing, or because the data represent a subjective interpretation and would therefore vary from one interpreter to another, or because measuring instruments are known to be imprecise, or for a host of other reasons. Thus rather than describe the magnitude of uncertainty, an error model permits its simulation. The best known error model is the Normal or Gaussian distribution, which simulates the possible values of a simple measurement by sampling a bell curve, and describes the magnitude of uncertainty as the distribution's standard deviation. The Circular Normal distribution, a two-dimensional version of the bell curve, is commonly used in spatial databases to describe the uncertainty in the position of a point's location, or as the basis of map accuracy standards.

In the context of spatial data, an error model provides a range of possible maps, each of which might represent the truth. The chapters in this section by

Fisher and by Goodchild et al. present simple error models for a particular class of spatial data, the chorochromatic or 'area class' type exemplified by soil or land cover maps, and use it to create a range of possible maps or realisations. Visually, the impact of one or more realisations may be much greater than the more abstract concept of an error descriptor, since the user is exposed directly to the range of possibilities available rather than to a measure of the range.

This distinction between visualization of error descriptors on the one hand and realisations of an error model on the other is illustrated somewhat dramatically by Englund (1993) in the context of Kriging estimation. When the technique of Kriging is used to create an interpolated surface between sample points of known value, the result is both a surface and a map of uncertainty. In fact the surface is the estimated mean, and the map of uncertainty shows estimated variance around the mean. Englund deviates from common practice by showing not the map of estimated means, but sample maps from the distribution of possibilities defined by the means and variances. These, rather than the estimated mean, are then used in GIS processing. As a result, Englund is able to provide visually dramatic illustrations of the uncertainty expressed by the estimated variances, but normally ignored in analyses based on estimated means.

... UNCERTAINTY IN THE DECISION-MAKING PROCESS

Uncertainty is an inevitable characteristic of spatial data, because it is impossible to measure positions on the Earth's surface to infinite precision, reality changes faster than maps, and a host of other common reasons. Regrettably, the traditions of map-making include remarkably few methods for displaying uncertainty. In part this is because it may not be visible at the scale of the map. But spatial databases have no scale, and force us to deal explicitly with uncertainty. In part also the lack of attention to uncertainty is attributable to the more aesthetic aspects of cartographic tradition. Uncertainty would devalue the map, and confuse its aim of communicating an interpretation of the landscape to its readers. However, in GIS, neither of these arguments is valid, and the lack of methods for visualizing uncertainty is a major drawback of systems intended to support objective analysis, scientific investigation and spatial decision-making.

We have little experience to date of spatial decision-making under uncertainty (Leung, 1988) or of the potential impact of visual displays of uncertainty on the process. Information on uncertainty is often available to map users in the form of text descriptions, but these are often published separately and may be hard to find. A visualization of uncertainty in a GIS would immediately draw the user's attention to a potentially problematic aspect of the project. Increasingly, digital maps and GIS are used for enforcement of regulations, particularly in environmental management, and direct access to the uncertainty inevitably present in the data could have dramatic effects on the regulatory process. Should maps of US wetlands derived by classifying remotely sensed images, and therefore inevitably subject to uncertainty, be used to enforce wetland regulations? Should imperfect maps of Spotted Owl habitat be used to regulate old growth logging in the Pacific Northwest?

145

The approach used to visualization, whether it be suppression of uncertainty, display of error descriptors, or display of alternative realisations of an error model, clearly will affect the decision-making process. Moreover its effects will differ depending on context. In scientific research, statistical methods can be used to estimate the uncertainty in a GIS product based on uncertain data. In a legal context, where a court is attempting to resolve conflict, the approach will be very different and based more on process than on demonstrable truth. A court may hold that a decision based on uncertain data is valid if it can be demonstrated that reasonable care and control was exercised over the creation and processing of the data, and provided that these meet agreed professional standards. In other words, the fact that it is impossible in principle to ensure perfect representation of the truth is not important. The court is concerned instead with whether adequate procedures were in place to ensure a reasonable level of accuracy. The role of visualization in this legal debate is far from clear.

... VISUALIZATION TOOLS FOR DISPLAYING VALIDITY

The degree to which uncertainty has been recorded will affect the way in which it can be visualized. Many spatial databases currently contain little or no embedded information on accuracy, certainty or other components of validity. Those data formats that do embed capabilities for storing metadata (for example, Vector Product Format) homogenise validity information across data layers or coverages. Spatial variation in validity may occur within data layers as a result of data collection, or may result from particular algorithmic processes applied to the data. Cressie (1991) provides an example produced as a by-product of Kriging.

In many cases, graphical displays of data have the potential to reveal the variation in spatial or temporal accuracy. Map animation should be of utility for cleaning any dataset characterised by high serial correlation. Spatial inconsistencies may also be detected efficiently by visual inspection, when graphical methods are chosen to bring inconsistencies to the attention of the viewer.

Propagation of error during GIS operations is expected, although specific models for monitoring and compensation have not been developed in all data domains (but see Fisher, 1991). In some cases, combining separate but interrelated data sources can reveal inconsistencies between them. For example, overlay of digitised vector transport networks over a rectified satellite image might point out conflation in the vector layer caused by digitising from multiple map sources. Of course, relative discrepancies do not indicate which data source contains inaccuracies in every case. And conversely, failure to detect errors by visual inspection does not indicate a clear absence of uncertainty.

Two modes of visualization can be applied to display the validity of GIS data. Traditional cartographic displays provide static snapshots of data surfaces and objects, affording visualization at specific or selected phases of data processing. For example, visual displays might be used to determine drifts and trends during iterative modelling and simulation. Dynamic visualization tools are emerging as an alternative mode for cartographic data display, and may also prove worthwhile for viewing and exploring information on data validity.

In the static mode, cartographic conventions are guided by application of Bertin's (1985) visual variables, including size, shape, texture, value, colour, orientation and arrangement. The human information processing system has strongest acuity for distinguishing symbol size, value (relative darkness or gray-tone) and colour (hue). With respect to validity, the visual variables seeming to offer best opportunities for display include value, colour (saturation or intensity) and texture. The visual metaphor of fog has been proposed (Beard, Buttenfield and Clapham, 1991) to alert viewers to uncertainty in data position or attribution. Fog may be created graphically by manipulating value or by de-saturating colours. MacEachren (Chapter 13) cites Woodward's idea of de-focusing a display for visual impact, which is an earlier presentation of a similar concept. Other visual variables such as texture may be used to indicate invalid data. For example, a coarse-texture screen may introduce visual noise as a metaphor for noisy data. The use of colour in graphical display is recognised in empirical research to be especially complicated (Brewer, 1989). Recent empirical research at NCGIA indicates that viewer recognition of the concept 'uncertainty' is not tied to linear progressions of either value or saturation in the RGB colour cube.

The visual variable Bertin refers to as 'arrangement' incorporates two-dimensional position and apparent three-dimensional position, simulated in a CRT display by means of hidden surface and hidden line removal, shading and size variation (distant objects appear to be smaller). The arrangement of objects in a display may be used equally well to imply logical inconsistencies in a data structure. For example, polygon fill can be used to check efficiently for closure of digitised polygons. Where topological inconsistencies occur, the fill tone or pattern will not match the polygon boundaries. Other graphical methods for strengthening or deconstructing the holism of image objects may also be applied, relying on many principles of Gestalt psychology. Viewers try to make sense of nonsensical data, interpreting compound geometric relations by visual isolation of regular shapes, and searching for familiar patterns in ambiguous visual fields. To make sense of an image, viewers will tend to formulate a mental model of what that image represents, and make comparisons to interpret the image. The greater the level of image abstraction, the more difficult the comparison. Application of visual variables to the display of abstractions about data validity probably requires development of skills beyond traditional map use exercises in current GIS curricula.

The alternative to static modes of display is called dynamic cartography by some, map animation by others, and is discussed in detail earlier in this book. Dynamic displays provide several advantages to viewers and designers of GIS displays, particularly in the context of validity and metadata. One can imagine a data sequence in which data and metadata displays are toggled, either automatically or by viewer choice; at some speeds, the data quality may appear to become embedded in the data display, while other speeds will isolate the two images. Another advantage is that for positional accuracy, symbol movement overcomes the problem of viewers attaching too much weight to particular cartographic representations. Fisher (Chapter 19) demonstrates this principle for soils maps.

Advances in display refresh speeds and CPU speed have greatly improved functionality for interactive graphics, and this can be especially useful in the context of data validity. Viewers can use point-and-click interface functions to query specific map objects in a display. One can imagine a hypermedia image in which clicking on a map object (a soil polygon, for example) will bring up a dialogue box, statistical chart or tabular display of lineage information to alert viewers to the date of data collection, scale of the data source, types of processing algorithms applied to generalise and symbolise the map image, and so forth. These capabilities lie somewhat beyond the scope of the current chapters, but there is every reason to expect that these types of functions can and will begin to be incorporated soon in upcoming versions of GIS software.

··· SUMMARY

This introduction has offered three levels on which one might approach issues and display of data validity: first, the development of frameworks linking spatial data quality components with visualization methods; second, the implementation of prototype displays; and third, error tracking and monitoring of error propagation during GIS operations. These interrelated concepts may be incorporated into a variety of measures of validity, or mapped using a variety of illustrative techniques. In Chapter 16, they are generalised into a triad of concepts: accuracy, resolution and consistency. Chapters 17 and 19 discuss multiple stochastic realisations of soils data and remotely sensed imagery, and present applications graphically. Chapter 18 by Jo Wood presents several methods to illustrate the validity of digital terrain data.

Research issues that remain to be addressed include refined methods for estimating error propagation rates, particularly in environmental monitoring where policy decisions are often based upon visual display, such as wetlands protection and climatic change studies. Additionally, the potential for cognitive overload in displays where data and data validity are embedded forms an important area for evaluation. Finally, the application of innovative graphical techniques to displays of validity has only recently begun to be addressed. Animation, multimedia and hypermedia may expand GIS viewer capabilities for exploration of data pattern and data validity. Considerations for design and evaluation of these media remain to be uncovered in further research.

··· KEY READING FOR THIS SECTION

Overviews of the data quality issue in spatial data can be found in M.F. Goodchild and S. Gopal (1989) *The Accuracy of Spatial Databases* (London: Taylor and Francis); in the chapter by Goodchild (1991b) in J C Muller (ed.), *Advances in Cartography* (New York: Elsevier, pp. 113–40); or in the chapters by Fisher and by Chrisman in D.J. Maguire, M.F. Goodchild and D.W. Rhind (eds) (1991), *Geographical Information Systems: Principles and Applications* (Harlow: Longman).

A useful discussion of the accuracy of map analysis techniques can be found in D.H. Maling (1989) *Measurements from Maps* (Oxford and New York: Pergamon). The Beard, Buttenfield and Clapham report listed in the references is a comprehensive overview of visualization of spatial data quality, and includes a prioritised research agenda.

...REFERENCES

Beard M K, Buttenfield B P, Clapham S (1991) *Visualizing the Quality of Spatial Information: Scientific Report of the Specialist Meeting.* National Center for Geographic Information and Analysis Technical Report 91–26, Santa Barbara, Calif.

Bertin J (1985) *Graphical Semiology,* University of Wisconsin Press, Madison, Wisconsin

Blakemore M (1984) Generalization and error in spatial databases. *Cartographica* 21(2–3): 131–139

Brewer C (1989) The development of process-printed Munsell charts for selecting map colors. *The American Cartographer* 16(4): 269–278

Cressie N A C (1991) *Statistics for Spatial Data,* Wiley, New York

Englund E (1993) Spatial simulation: environmental applications. In: Goodchild M F, Parks B O, Steyaert L T (eds) *Environmental Modeling with GIS,* Oxford University Press, New York

Fisher P F (1991a) First experiments in viewshed uncertainty: the accuracy of the viewshed area. *Photogrammetric Engineering and Remote Sensing* 57(10): 1321–1327

Goodchild M F (1991) Issues of quality and uncertainty. In: Muller J-C (ed.) *Advances in Cartography,* Elsevier, New York, pp. 113–140

Leung Y (1988) *Spatial Analysis and Planning under Imprecision,* Elsevier, New York

US Geological Survey (1987) *Digital Elevation Models – Data Users Guide,* US Geological Survey, Reston, Virginia

16 GRAPHICAL AND GEOGRAPHICAL COMPONENTS OF DATA QUALITY

B. BUTTENFIELD & M. KATE BEARD

... INTRODUCTION

New hardware and software currently allow fast processing and display of large volumes of information. National, regional and local spatial databases are accumulating in many countries to support a wide range of applications. With development and dissemination of these databases, users become increasingly removed from the details of data collection and processing. Intimacy with the data and an understanding of its limitations are lost. Computer generated maps, a standard output of a GIS, generally imply an accuracy not warranted by the data. Misinterpretation of GIS results can lead to false claims, poor judgements and even litigation (Epstein and Roitman, 1987). There is a sense of urgency within the GIS community to address these problems by developing capabilities to accumulate, store and communicate information describing the quality of the data in the database (Lanter and Veregin, 1990).

Information on the quality of spatial data and databases has formed a long-term concern for both developers and users of GIS (Chrisman, 1983). Data producers are concerned about the utility and credibility of digital products. Users are concerned about the reliability of interpretations and decisions based on such products. Tools for the visualization of data quality have not been well developed and are not readily available in existing GIS packages. The volume of information required to describe spatial data quality is potentially quite large, and this presents a challenge for GIS development.

This chapter reports development of a conceptual framework within which to address research topics associated with visualization of data quality. The framework refines previous work (Buttenfield and Beard, 1991). A brief review of other efforts to report data quality prefaces discussion about the framework.

... ALTERNATIVES FOR REPORTING DATA QUALITY COMPONENTS

Defining data quality components is the first step towards determining which aspects can be visualized effectively. One commonly cited component of data quality is error. Recognised types of error include those associated with data collection, called source error, but Beard (1989) defines a second component, use error, that is associated with the appropriate application of data or data products for specific applications. A third type, known as process error, has proved difficult to formalise, for example in modelling the error associated with soil mapping (Fisher, 1991). Statistical models often incorporate error terms, but few capabilities for analysing error appear in current statistical packages. Sensitivity to error varies with the type of data and use. One tends to overlook error in census data because of the nature of its collection and processing. One tends to anticipate error in a soil map, whose quality often improves with additional money and sampling resources. However, frameworks based solely on error do not allow for other dimensions of data quality, such as resolution or consistency.

Efforts to arrive at a broader definition of quality are progressing in the United States. A coalition of representatives from the public and private sector is collaborating to facilitate transfer of digital datasets between federal, state and local agencies. The preliminary result was a proposal for a standard introducing five categories of data quality (Moellering, 1988). Subsequent refinements are incorporated in the Spatial Data Transfer Standard (SDTS, 1990). Federal producers of digital data currently reviewing the standard include the US Bureau of the Census, the US Geological Survey, and the US Defense Mapping Agency. Other nations, notably France and Canada, have undertaken similar efforts, and the SDTS is presented here as an example, not as the ultimate solution.

Categories for data quality in the SDTS include measures of positional and attribute accuracy, logical consistency, data completeness and lineage. Positional accuracy generally refers to horizontal and vertical control within a geodetic coordinate system (e.g. latitude–longitude), although statistical coordinate systems may be appropriate (e.g. principal components factor scores). Attribute accuracy refers to errors associated with categorisation or classification of thematic variables. Completeness relates to issues of missing data and methods of compensating for missing values. Logical consistency applies to the data structure, and checks for contradictory relations in the database. Lineage implies a chronological record of data sources and applied processing operations.

While the SDTS provides guidance on reporting mechanisms for datasets as a whole, it does not distinguish between geographic and digital phenomena. An example in transportation may clarify this point. Concern with the geographical phenomenon might indicate whether road conditions on a particular highway are wet or dry, foggy or clear. Uncertainty in the digital realm might concern whether or not the road category is correctly classified as a divided highway (dual carriageway), or correctly attributed as lying within federal or provincial jurisdiction. There is an important distinction between uncertainty about real-world conditions (validity of geographical knowledge) and uncertainty about database descriptions (validity of data). SDTS require data validity reports at the time of data production. The standard does not address geographical validity, as for example, by mandating periodic updates of quality reports.

A second factor limiting SDTS is that it does not provide for variations in quality based on the types of objects stored in each digital dataset. Quality information may vary dramatically, because of differences in digital representation or differences in application. For example, objects stored in soils database formats include two kinds of geometric objects, called field check points and soil parcels. Linear objects are derived, not collected, and they are meaningful only in the digital domain. Polygons are derived by interpolation and overlay operations. Conversely, current versions of vector database formats (such as DLG-E used at US Geological Survey), provide for point, line and polygon objects. Points may have digital significance (such as the 'nodes' required to generate polygons, and 'entity points' such as towns or benchmarks). Polygon objects are digital artifacts, generated by topological operations.

Upon reflection, one may generalise three components of data quality: accuracy, resolution and consistency. Accuracy differs conceptually from error in measuring discrepancy from a model; error measures discrepancy from the truth (Buttenfield, 1991; Clapham, 1992). The two concepts are not converses of each other. Measurement tends to impose models upon the things being measured, and this is particularly true in the social sciences. 'True' values are not directly observable, which is why standards such as SDTS define measures of accuracy instead of error. Resolution includes measurement or inference of either spatial or temporal elements. Resolution includes data completeness, in the sense that the proportion of missing values may compromise data resolution overall. Consistency typically refers to compliance with topological relations (internal consistency) or to compliance with attribute codes and attribute hierarchies (external consistency).

A framework incorporating the variety and realism of observable geographic phenomena with generalisable data quality components will overcome some of the SDTS limitations. The framework must be flexible, to accommodate various digital data models and data objects, and must facilitate application of visual tools for generating graphical representations of information about quality.

... INCORPORATING GEOGRAPHIC REALISM INTO DATA QUALITY RECORDS

Sinton (1978) models location, theme and time as required plus interdependent observation elements. In spatial data collection, one of these elements is typically fixed, another is controlled, and the third is measured. In demographic sampling, the time (usually a five-year or ten-year census) is fixed, enumeration units are controlled (location), and demographic attributes (theme) are measured. Data quality restrictions will vary accordingly. In the census, the controlled component (location within enumeration districts) may be associated with errors due to aggregation levels (Fotheringham and Wong, 1991). The fixed component (date of the census) may distort the currency of the representation. These errors differ from error associated with the measured attribute or theme (for example, age, income level, or ethnic background), which may incorporate both data collection error (source error) and error inherent in the classification algorithms applied to the data (process error).

For any geographical observation, locational, thematic and temporal elements can be cross-tabulated or sampled for varying levels of data quality as shown in Figure 16.1. Accuracy measures may record locational, thematic and/or temporal indices. Consistency may incorporate the logic of the data structure as well as the uniformity of sampling frame (positional consistency), the logic of categorisation (thematic consistency), or the sampling periodicity

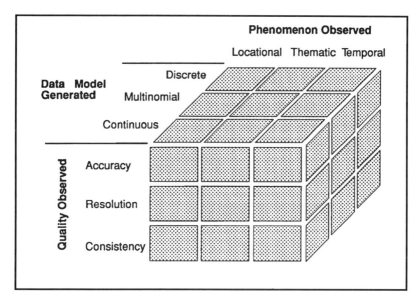

Figure 16.1 Components of spatial data quality observed with spatial information

(temporal consistency). Queries to a database to select and map all data updated in the past six months whose position lies within first-order geodetic control formulate a reasonable and realistic request for quality information. The matrix framework can handle a specific request such as given in this example. Additionally, modular organisation of the framework supports efficient query. Many queries will not require access to all matrix cells to formulate a response.

Other examples occur in the context of geographical observation and analysis. Root mean square error commonly describes locational accuracy. An important attribute of satellite imagery is the sensor's digital resolution. Tobler (1987) measures attribute resolution of census data by computing the average size of enumeration units. Temporal consistency is a high priority for climatological summaries, where slight variations in the hour of data observation can affect daily and monthly summaries substantially. Measures of thematic accuracy include covariance of a set of observations, or proportions of soil inclusions. Compliance of thematic relationships should require consistent application of rules, for example that all highway segments require consistent feature codes, and so on.

... INCORPORATING GRAPHICAL LOGIC INTO THE FRAMEWORK

The third dimension of Figure 16.1 integrates data quality measures with the type of data model applied. Particular GIS operations generate a particular type of data model, and are designed to organise particular data objects. Discrete data models represent individual objects with clear spatial or temporal bounds, derived features such as network links and nodes, and tangible geographical features (streets, cities, etc.). Quality measures for discrete models include error ellipses and other discrete symbols. Polygon-based data models organise categorical data generated by overlay (as for soils or land use) or by partitioning (as for metric classed data such as population density). Quality displays for categorical data include misclassification matrices or covariance matrices, for example. Solid models and/or the surfaces bounding solids represent continuous data generated by interpolation, as for elevation, temperature or density observations. Volumetric measures associate data quality with continuous data, and include autocorrelation measures, or maps of residuals.

Incorporating data models into the framework provides flexibility for digital representation of many data types. The framework also provides for appropriate graphical representation. The symbols encoding the graphical display should be logical with respect to the data model. The visual variables proposed by Bertin (1985) integrate a consistent mapping between nominal, ordered and metric relations and graphic encoding schemes. Visual variables include size, shape, texture, value, colour and orientation. The strongest acuity in human visual discriminatory power relates to varying size, value and colour. These visual variables often represent quantitative data types. Manipulation of colour (hue) and geometric shape illustrate nominal or qualitative distinctions. Symbol orientation and visual texture do not by themselves provide reliable visual distinctions. These two sometimes augment or emphasize the message implied by other visual variables.

To retain flexibility for queries about the data quality information, the framework must relate elements of data quality back to the type of phenomena observed. To visualize the data quality information, links between the data model and the display should be clear and non-ambiguous. This is a two-step process. The graphical display does not represent the observed phenomena, but rather its digital description, based on a selected data model. The map feature is not the same as the geographical phenomenon it represents. The data model guides and constrains the design of the display.

... GRAPHICAL CONSIDERATIONS FOR FURTHER RESEARCH

Priorities for research relate to the design and implementation of specific visualization techniques for representing data quality. In exploring design decisions, several classes of research questions arise. These include issues of scale, contextual dependency, data modelling issues, and maintenance of links between digital representation and graphical depiction.

In GIS the representation scale is dynamic and the appearance of many features varies with scale. Quality components linked to scale-dependent features and attributes may be dynamic as well. This is particularly true of locational accuracy and resolution, whose visual representations may need to vary with scale. The scale of data display may be incompatible with a meaningful display scale for data quality. Error ellipses based on 10 metre resolution may be indecipherable on a map of 1:100 000, and little is known about the cognitive implications of embedding multiple scales within a single map fabric. Additionally the grain of two different quality measures may differ by orders of magnitude, making it difficult to embed both in the same display. This raises the issue of whether GIS display capabilities can support local and global views simultaneously and intelligently. For example, if a quality component is insignificant at a particular scale, should system graphical defaults render it non-visible? Or should this decision be left to a (possibly) unsuspecting user?

Examination of quality information will vary in rigour and level of detail for specific GIS applications. Computationally expensive visualization techniques may not be appropriate for certain tasks. In the case of digital data production, visualization techniques support a very narrowly defined set of quality control parameters for a specific data product. Display design can accordingly be specific. The graphical display might encode a simple colour scheme that quickly indicates compliance with tolerance thresholds. In a spatial decision support context, requirements for quality information are more broad. For example in responding to a fire, positional accuracy may be irrelevant but attribute accuracy on the age and structure of buildings, presence of hazardous contents, and currency of digital data may be critical. In the context of spatial inference, requirements for data quality information are neither simple, consistent nor pre-defined. In the disproof of hypotheses, the quality of the data must be rigorously established and tracked. Quantitative and qualitative assessments of quality measures will likely require visualization techniques supporting metric analysis. The design of the graphics in each case should meet requirements for quality information. There are implications for system design as well as for visual perception and cognition.

The error in raw data and error introduced during data manipulation is not always controllable, and for many GIS operations there is no direct way to obtain accuracy measures. A particular statistical problem with spatial data is that replications of GIS models are not typical. Another computational issue centres on the lack of robust models describing error propagation during GIS operations. Without formalised expectations, it is difficult to monitor or predict the generalised behaviour of analytical procedures. Empirical evaluation of graphical tools to monitor error propagation forms an important direction for future research.

What data models lend themselves to representations of data quality? Hierarchical models similar to the quaternary triangular mesh (QTM) structure developed by Dutton (1989, 1992) may be useful for some quality indices. If the quality of information varies spatially, it should be possible to decompose the area into QTM triangles until each triangle has a uniform level of quality. Automatic system warnings could alert users (for example) attempting to make inferences beyond the usable limits of resolution defined by homogeneity in the QTM hierarchy.

Should the graphic representation embed the same symbol with data and data quality information, creating a multivariate encoding? Preliminary work at SUNY Buffalo indicates that multivariate schemes require viewer training. In an experimental animated map, multivariate symbols illustrate annual and cumulative tree-ring growth. The cumulative growth represents a running tally of the trunk size at various points in time. The annual growth represents a quality value associated with the vigour and intensity of growth year by year. (In the framework of Figure 16.1, annual growth provides an example of thematic consistency.) The animation shows cumulative growth by changing symbol size, and shows intensity of growth by varying the intensity (saturation) of symbol colour. Viewers of the animation do associate gradations in symbol size with the cumulative growth values, but tend to ignore or be confused by gradations in colour saturation or value embedded in the same symbol. Without some prior indication that the saturation indicates the quality of growth, some viewers identified colour saturation with changing seasons. Clearly the visualization of data quality requires careful graphical design coupled with evaluatory assessment to be certain that displays are logical and unambiguous.

··· SUMMARY

This chapter presents a framework linking observation of geographic phenomena to data quality measures using linkages with particular types of data models. The goal is to incorporate geographical realism and graphical logic into data quality representations. Decomposing data quality into components of accuracy, resolution and consistency allows data producers to isolate the components of primary concern and to track them separately. In many cases these quality components are not independent and an understanding of the relationships between them becomes important during database query.

Questions for future research include study of the relationships between the three quality components, and formalisation of guidelines to match appropriate data models with specific quality measures. Refinement of visual tools to monitor error propagation during GIS modelling forms a second arena for valuable research. Another research priority concerns evaluation of visualization techniques to assure that they match user expectations, and do not exceed the limits of perceptive and cognitive skills. Without a clear sense of what GIS users expect to discover in a data quality display, or even how users comprehend data quality concepts and implications, software and database solutions cannot be optimised. Discovery and innovation have traditionally involved visual thinking. This argues for the importance of understanding information processing capabilities that rely on visual tools and skills. Evaluation of visual solutions to data quality representation requires attention to the perceptual and cognitive mechanisms by which spatial and temporal patterns are interpreted.

··· ACKNOWLEDGEMENTS

This research forms a part of Research Initiative 7, Visualizing the Quality of Spatial Information, of the National Center for Geographic Information and Analysis (NCGIA). Funding support for NCGIA provided by the US National Science Foundation (NSF SES 88–10916) is gratefully acknowledged. Provision of travel funding to the workshop by the Association for Geographic Information is also acknowledged.

··· REFERENCES

Beard M K (1989) Use error: the neglected error component. *Proceedings AUTO-CARTO 9*, Baltimore, Maryland, pp. 808–817

Bertin J (1985) *Graphical Semiology*, University of Wisconsin Press, Madison Wisconsin

Buttenfield B P (1991) Visualizing cartographic metadata. In: Beard M K, Buttenfield B P, Clapham S (1991) *Visualizing the Quality of Spatial Information, Scientific Report of the Specialist Meeting*, NCGIA Technical Report 91–26 Santa Barbara, California

Buttenfield B P, Beard M K (1991) Visualizing the quality of spatial information. *Proceedings AUTO-CARTO 10*, Baltimore, Maryland 6: 423–427

Chrisman N R (1983) The role of quality information in the long-term functioning of a geographic information system. *Cartographica* 21 (2–3): 79–87

Clapham S (1992) *A formal approach to the visualization of spatial data quality*. Unpublished MSc Thesis, Department of Surveying Engineering, University of Maine.

Dutton G (1989) Modelling locational uncertainty via hierarchical tesselation. In: Goodchild M F, Gopal S (eds) *Accuracy of Spatial Databases*, pp. 125–140

Dutton G (1992) Handling positional uncertainty in spatial databases. *Proceedings, 5th Spatial Data Handling Symposium*, Charleston, South Carolina, 2: 460–469

Epstein E, Roitman H (1987) Liability for information. *Proceedings, Annual Meeting of the Urban and Regional Information Systems Association* (URISA), pp. 115–125

Fisher P (1991b) Modelling soil map-unit inclusions by Monte Carlo simulation. *International Journal of Geographical Information Systems* 5(2): 193–208

Fotheringham A S, Wong D W S (1991) The modifiable areal unit problem in multivariate statistical analysis. *Environment and Planning A*, 23(7): 1025–1044

Lanter D P, Veregin H (1990) A lineage meta-database program for propagating error in geographic information systems. *Proceedings GIS/LIS 90*, Anaheim California, 1: 144–153

Moellering H (ed.) (1988) The proposed standard for digital cartographic data: report of the digital cartographic data standards task force. *The American Cartographer* 15(1)

SDTS (1990) *The Spatial Data Transfer Standard*. US Government Printing Office, US Geological Survey Bulletin, Washington DC

Sinton D (1978) The inherent structure of information as a constraint to analysis. In: Dutton D (ed.) *Harvard Papers on Geographic Information Systems*, Addison Wesley, Reading.

Tobler W (1987) Measuring spatial resolution. *Proceedings International Workshop on Geographical Information Systems*, Beijing, pp. 42–48

17 Visualizing fuzzy maps

M. Goodchild, L. Chih-Chang & Y. Leung

... Introduction

Increasing emphasis on analysis, modelling and decision support within the GIS applications community in recent years has led to a general concern for issues of data quality. If the purpose of spatial data handling is to make maps, then perhaps it is sufficient to require merely that the output map product be as accurate as the input and to present visualizations of the data as if they were perfectly accurate and certain. But the detailed analytic and modelling applications that underlie much of the recent literature of GIS (Tomlin, 1991; Laurini and Thompson, 1992) demand much more stringent and robust approaches. If the input is known to be inaccurate, uncertain or error-prone, then it is important that the effects of such inaccuracies on the output also be known. Without such knowledge, the apparent value of GIS in supporting spatial decision-making may be illusory. Similarly, it is important that the user be fully informed about data quality, and the most effective way to do this may be by some method of visual communication. In this sense, the arguments for visualization of data quality in GIS seem much stronger than in cartography, and perhaps explain why there are so few techniques for data quality display in the cartographic tradition.

As noted in Chapter 15, uncertainty in spatial databases has many sources, not all of which are captured by the terms 'error' and 'inaccuracy'. The contents of the database can differ from the truth, or from some source of higher accuracy, because of the effects of abstraction in the map-making process, through generalization, abstraction, exaggeration, simplification or classification. All of these create forms of uncertainty for the analyst, and in principle it is desirable to be able to measure, or at least perceive, their effects.

Although inaccuracy is pervasive in spatial data, some types of data are clearly less accurate than others. A GPS (Global Positioning System) survey provides known levels of positional accuracy, potentially down to the nearest centimetre. We focus in this chapter on a class of data known to be subject to relatively high levels of uncertainty, and for which there are no such straightforward measures of accuracy. In this class, every point on the plane is characterised by a single value measured on a nominal or multinomial scale; examples include soil class, land cover class and land use. We refer to this as a multinomial field and such fields are often displayed by chorochromatic, or k-colour, maps. Two data models are commonly used to build digital representations of such fields. The first, the raster model, is used when the field is obtained by remote sensing, by making use of one of a number of standard procedures for classification. In this model, spatial variation is represented through an array of rectangular cells or pixels, and all information on within-pixel variability is lost. The second, or polygon model, partitions the plane into a number of polygons of arbitrary shape but homogeneous class, thus losing all variability within polygons. The polygon model is also commonly used in making maps of multinomial fields, although the boundary lines on such maps are drawn as continuous curves, and will be discretised as polygons, with straight line segments between vertices, only when the lines are digitised.

Both models are clearly approximations. In the raster model, all variation over distances less than the cell size is lost, and cell size is therefore a convenient descriptor of positional accuracy. The accuracy of the map as a whole is also determined in part by the selection of classes, since in reality, soils or land cover variation is only approximately described by a fixed, finite number of classes, and there will always be within-class variation in the real world which the map purports to represent. In the polygon model, variation within polygons is lost, but polygons are not fixed in size and so there is no simple measure of positional accuracy to compare to the cell size of the raster model. In practice, positional accuracy is often quoted as boundary width, typically 0.5mm, but although boundaries may be drawn with thin lines, the actual width of transition zones in reality may be much larger, and such measures fail to capture the positional uncertainty attributable to a lack of homogeneity within polygons. A better alternative is to derive a measure of positional accuracy from the area of the smallest mapped polygon, or 'minimum mapping unit', since patches smaller than this must have been ignored. A suitable measure is the diameter of a circle of area equal to the minimum mapping unit.

By itself, a measure of positional accuracy does little to capture the inherent uncertainty in raster or polygon representations of multinomial fields. It fails to deal with the problem of within-class variability, or variation over distances less than the positional accuracy of the data. Recently, GIS researchers have begun to turn to concepts of fuzzy classification to provide a more versatile approach to characterising uncertainty, and fuzzy classifiers have become popular sources for error descriptors and error models. The objective of this chapter is to illustrate a number of methods for visualizing this particular approach to uncertainty in multinomial fields. In the raster case, fuzzy classifiers provide a means of describing uncertainty, by associating each pixel not with a single class, but with a vector of class memberships, each one interpreted as a measure of belonging. Thus pixel x's degree of belonging in class i might be denoted by $\pi(x)$, and the vector of class memberships might be written:

$$\{\pi_i(x), \pi_2(x),..., \pi_n(x)\}$$

where n is the number of classes.

In the polygon case, inaccuracy occurs in the form of variation within polygons, perhaps at the edges where boundaries are merely approximations to zones of transition (Mark and Csillag, 1989), or perhaps centrally where small

159

inclusions and islands of different classes have not been mapped because they fall below the minimum mapping unit area. Neither of these issues is dealt with effectively by giving the polygon a fuzzy class membership. Instead, it is necessary to abandon the polygon model because it is fundamentally unable to serve as an adequate basis for representing within-polygon variation. Instead, we see the geometry of the polygon model as an artifact of the mapping process, having little value in an effective approach to data quality, and transform to the raster model. Thus both heterogeneity of polygon class and transition near the boundary are represented through the use of pixel class memberships.

While the concept of fuzzy pixel classification is a familiar feature of the remote sensing literature, there has been very little research on its visualization, or on the processing of such data within GIS. In part this may be because of concerns over data volume, since n memberships must be stored for each pixel, rather than one integer between 1 and n. In practice, however, it is rare for more than two class memberships to be significantly greater than zero in any one pixel. Fuzzy-classified scenes are difficult to visualize for similar reasons, because the objective in principle would be to communicate n memberships to the user per pixel. Moreover it is not clear how measurements such as class area can be made from such data. Thus despite the availability of fuzzy classifiers, and the greater information content of fuzzy-classified scenes, it is tempting to convert such data to a simple maximum likelihood classification on the grounds that the latter are much easier to handle. As a result, estimates of the area of a given class are biased, and the user viewing a maximum likelihood display is given a falsely optimistic impression of successful classification.

In cartography, a polygon with fuzzy or mixed classification is sometimes shown filled with bars of alternating colours, corresponding to the two classes that are mixed in the polygon. The use of bars of constant width and straight parallel sides allows the eye to distinguish correctly between true boundaries, which have generally complex shapes, and the artificial boundaries formed by the bars. However the method is effectively limited to mixtures of two classes, and becomes confusing if more than a small proportion of polygons are mixed, and there is still the risk that the uninformed user will misunderstand the convention.

The purpose of this chapter is to discuss methods of visualization and processing for fuzzy-classified maps and scenes within GIS. We include with this term not only the results of fuzzy classification in remote sensing, but also derivatives of the polygon model where each pixel is associated with a mixture of classes, or with probabilities of class membership. The next section discusses the meaning of such data from a statistical perspective, and introduces the chapter's error model. This is followed by a description of the environment for visualization of fuzzy-classified scenes developed by the authors. The final summary discusses directions for future research. Sections of the chapter draw on Leung, Goodchild and Lin (1992).

...PROBABILISTIC PERSPECTIVE

Consider a raster in which each pixel is associated with a vector of class memberships. The various possible sources of this data were discussed in the previous section. To provide a probabilistic interpretation, we assume that the memberships are normalized by pixel:

$$P_i(x) = \frac{\pi_i(x)}{\Sigma_k \pi_k(x)}$$

Thus $P_i(x)$ is interpreted as the probability that pixel x belongs to class i out of the n classes. This might be understood in a mixed pixel context as the proportion of pixel x's area that is of class i; or the proportion of interpreters who would have assigned the pixel's area to class i; or the proportion of pixels with the same spectral response as x that are truly i; or in a seasonal sense as the proportion of the year during which the pixel should be assigned to class i rather than some other class. Numerous other interpretations are possible.

We define the term multinomial probability field (MPF) as a vector field whose value at any point is a normalised vector of class membership probabilities of length n. In other words, the database can be queried to determine, for any point (x,y), the probabilities that point belongs to any of classes 1 through n. A raster provides a suitable way of creating an acceptable approximation of such a field in a digital database, with a spatial resolution defined by the cell size.

Although a display of pixels showing the membership in each class is informative, it nevertheless fails to convey an impression of uncertainty, suggesting that memberships are expressions of deterministic knowledge, rather than of lack of knowledge, or of fuzziness. Instead, we focus on class memberships as descriptions of uncertainty, and on the range of possible maps that might therefore exist. In other words, and using the terms defined in the introduction to this section, the fuzzy class memberships become parameters of an error model, and the range of possibilities is defined by realisations of that model. Goodchild, Sun and Yang (1992) define an error model in the context of spatial databases as 'a stochastic process capable of generating distorted versions of the same reality'. The best known error model is the Gaussian, used to describe uncertainty in measurements of a simple scalar quantity like the elevation at a point. Each of the outcomes of such an error model provides one possible version of the truth, as it might be interpreted by one soil scientist, or as it might be digitised by one operator. Thus we see a map as a collection of interdependent measurements, and an error model for maps as a means of generating simulations of alternative maps within the error model's inherent range of uncertainty in those measurements. One realisation of a map error model might be one scientist's interpretation of variation in land cover over a given area, or one digitiser operator's effort to capture the contents of a given map.

Goodchild, Sun and Yang (1992) describe an error model for an MPF. Each realisation is a map in which each pixel is assigned to a single class. The model's two essential properties are:

(a) between realisations, the proportion of times pixel x is assigned to class i approaches $P_i(x)$ as the number of realisations becomes large; and
(b) within realisations, the outcomes in neighbouring pixels are correlated, the degree of correlation being controlled by a spatial dependence parameter ρ.

When the spatial dependence parameter is zero, outcomes are independent in each pixel. This is the case illustrated by Fisher (1991b). However, this is almost certainly unrealistic since few if any real processes are likely to create such independent outcomes. As the parameter increases, outcomes are correlated over longer and longer distances. One suitable interpretation of this is that larger and larger inclusions within polygons are ignored, or fall below the minimum mapping unit area.

For example, consider an agricultural field of 100ha, captured as a raster database with a cell size of 0.01ha (10m square). A soil map of the field might include four classes, with significant uncertainty because of fuzzy boundaries between classes, inclusions, etc. Because the pixel size of 1ha has no connection with any process of soil formation or development, it is next to impossible that the true class will be independent in neighbouring pixels. Instead, we expect neighbouring pixels to have strongly correlated classes, and inclusions to extend over distances substantially more than 10m. On the other hand, while there may be uncertainty over the crop grown in the field, it is certain that the field has only one crop. Spatial dependence is so strong in this case that in any realisation of the error model, the entire agricultural field can have only one class.

Many commonly used descriptions of map error fail to meet the requirements of an error model, since they fall short of the complete specification of a stochastic process. Such descriptions include the width of an epsilon band, the measures mandated by many map accuracy standards, the statistics of the misclassification matrix used in remote sensing, and the reliability diagram found on many topographic maps. All of these are useful error descriptors, but fall short of being useful error models. Neither is there a useful connection between many such descriptors and the necessary parameters of error models. For example, it is not possible to connect the parameters in the model described above with such measures as positional accuracy of polygon boundaries, or per-polygon misclassification of attributes. Visualization is perhaps the only way of overcoming this conceptual barrier, since it allows us to connect the model and its parameters with their implications in the form of simple pictures.

... TOOLS FOR VISUALIZATION

In this section we describe the tools we have developed for visualization of MPFs. These techniques include some designed to display the descriptors of error (parameters of the error model) and others designed to generate and display realisations of the error model. As we argued in the first section, classification procedures are important for remotely sensed imagery, but it is also desirable to be able to visualize MPFs from sources such as land cover maps in which the classification is performed by other means. For this reason, the system is modular in design, and includes a classification module, display module, and modules for data manipulation. Only the display modules are discussed here (for additional details see Leung, Goodchild and Lin, 1992). The system is interactive and uses a graphic user interface, all instructions and operations being triggered by selecting appropriate screen buttons. Windows are opened and closed as appropriate. It has been developed in C and X Windows for the IBM RS/6000 under the AIX operating system.

In the display module, images can be displayed directly by associating colours with spectral bands without classification, in order to support direct visualization of the preclassified scene. However the most important component of the module supports the display of classified images. In general, techniques of dithering and bit-mapping can be used to display uncertainty in terms of levels of class membership, to expose the spatial variation in membership within regions or across region boundaries. In addition the system provides several other measures and methods for conveying information about an MPF to the user. The following sections briefly describe and illustrate the principal tools.

The illustrations, Plates 17 to 20, are derived from a Landsat TM scene for an area to the west of Santa Barbara, California. For the purpose of illustration, the scene was classified using only three classes, identified as water, vegetation and soil. Since these clearly do not characterise the full range of spectral responses, pixel fuzzy memberships tend to be relatively low and mixed. This somewhat artificial classification example produces artifacts which are the subject of comment below.

UNCLASSIFIED IMAGE

Colours can be assigned to spectral bands to create conventional false-colour representations of the unclassified scene. This allows the user to see the raw data before classification.

CLASSIFIED IMAGE

The RGB (red/green/blue) colour model is used to display the results generated by the fuzzy classifier, or input from some other source. Each class is associated with a point in RGB space, and each vector of class memberships is mapped to an intermediate point in the colour space by linear interpolation. This method is successful for two classes ($n=2$) provided the pure-class colours are chosen carefully, but it is difficult for the eye to decode the results for $n=3$, and for $n>3$ the mapping from class membership vector to colour space is no longer unique. Moreover mapping is non-unique for $n=3$ if the class memberships have not been normalized to sum to 1 (see previous section).

Plate 17 shows a display of the image using this approach assigning water to blue, vegetation to green and soil to red. Although all of the information contained in the fuzzy memberships is displayed in this rendering, it is virtually impossible for the eye and brain to deconvolute the linear mixing of colours. On the other hand, this may be a useful display if the user merely wishes to identify the memberships associated with a given pixel; the bars at the top of the image display the RGB components of the pixel currently selected by the cursor.

To deal with the difficulty of visualizing membership in many classes, it is possible to display each class's memberships separately using a grey scale. By using multiple windows one can display the general distribution of each class for up to four or even six classes simultaneously. Plate 18 shows the memberships for the three classes water, vegetation and soil in the upper left, lower left

163

and upper right windows respectively. While these are easier to interpret than Plate 17, the displays convey no collective sense of a pixel's memberships.

Sometimes it is desirable to have a non-fuzzy image of a fuzzy scene. A simple defuzzing mechanism is maximum likelihood, where the displayed class $f(x)=i$ if $\pi_i(x) > \pi_j(x)$ for all i, j, i not equal to j; that is, a pixel is assigned to class i (and displayed with class i's colour) if its degree of membership in class i is highest of all its memberships. The user has control over the colours assigned to each class. Frequency distributions of the entire image can be displayed, and the user can zoom into a selected area, or display the contents of any pixel.

AREA

Calculation of the area occupied by each class is a common GIS function. For conventionally classified scenes or other forms of raster data it is calculated by counting the pixels assigned to each class and multiplying by pixel area. However the solution is less clear in the case of fuzzy-classified scenes. If $P_i(x)$ is interpreted as the proportion of pixel x that is truly class i, as in a mixed pixel interpretation of fuzziness, then the area of class i will be the sum of such fractions added over the scene. On the other hand if $P_i(x)$ is interpreted probabilistically, the same estimate must be interpreted as the expected area of class i. Similar approaches are appropriate if $P_i(x)$ is given other probabilistic interpretations. Thus the calculation of area on a fuzzy-classified scene seems adequately addressed by calculating:

$$A_i = b\sum_x P_i(x)$$

where b is the area of each raster cell. Note that this estimate may be very different from the conventional one based on maximum likelihood, that is,

$$A_i^* = b\sum_x f(x)$$

More difficult is the estimation of error variance, standard error, or the uncertainty associated with such estimates. In the mixed pixel interpretation A_i is deterministic, with zero uncertainty. In a probabilistic interpretation, and assuming that outcomes in each pixel are independent of outcomes in neighbouring pixels (zero spatial dependence) then the uncertainty associated with area estimates can be determined from the statistics of the binomial distribution in the form of a standard error:

$$e_i = b(\Sigma_x P_i(x)[1 - P_i(x)])^{0.5}$$

where e_i is the root mean square uncertainty in estimate A_i. Fisher (1991) used Monte Carlo simulation to estimate this standard error. When spatial dependence is present, as it almost always is, and outcomes in neighbouring pixels are correlated, it is necessary to resort to the methods described by Goodchild, Sun and Yang (1992).

ENTROPY

The degree of certainty in a pixel's classification can be measured in various ways, but one that expresses the degree to which membership is concentrated in a particular class, rather than spread over a number of classes, is the information statistic or entropy measure:

$$H(x) = \frac{1}{\log_e n} \sum_i P_i(x) \log_e P_i(x)$$

where $H(x)$ is the entropy associated with pixel x. $H(x)$ varies from 0 (one class has probability 1, all others have probability 0) to 1 (all classes have probability equal to $1/n$). The system allows a map of H to be displayed using a grey scale; light areas have high certainty (probability concentrated in one class) while dark areas have low certainty.

The distribution of the entropy statistic for the Santa Barbara scene is shown in Plate 19. Clearly evident is the ocean turbidity off Point Arguello, where mixing of currents produces substantial reduction in the membership of the water class, and thus an increasing level of entropy in these pixels. In general the ocean is dark because of its high membership in the water class, but the land pixels are much more mixed and thus lighter in this rendering. Some of the greatest levels of uncertainty are in pixels located in the Town of Lompoc, because urban is not a recognised class.

The degree of fuzziness associated with membership in each class can be assessed by another form of the entropy measure:

$$H_i = \frac{1}{N \log_e 2} \sum_x \{P_i(x) \log_e P_i(x) + [1 - P_i(x)] \log_e [1 - P_i(x)]\}$$

where the sum is now over the pixels and N is the number of pixels. Entropy is zero if the probability of membership in class i is 0 or 1 in all pixels, and 1 if probability is 0.5 in all pixels. The overall entropy H of the entire fuzzy scene can be obtained by adding these measures over all classes.

REALISATIONS

As noted earlier, an important aspect of visualizing uncertainty is the ability to view individual realisations of an error model, rather than its parameters. All of the previously noted methods display some aspect of the probability vectors, which are the parameters of the error model's stochastic process, rather than its outcomes. Viewing a display of probability vectors necessarily diverts attention from the variation between realisations, and focuses more on the average or expected case.

The system includes the ability to display realisations of the error model, using user-determined levels of spatial dependence. Goodchild, Sun and Yang (1992) discuss possible methods for determining appropriate levels, as attributes of the entire map, or of individual classes, or of geographic regions. A display of four or six different realisations in different windows on the screen provides

165

graphic illustration of the implications of uncertainty in spatial data, and draws attention to its influence on analysis, modelling and decision-making.

Plate 20 shows a series of realisations of the Santa Barbara scene, using probabilities computed by normalising class memberships. Each of the four images shows a different value of spatial dependence ρ, ranging from 0.20 in the upper left to 0.25 in the lower right. In reality, spatial dependence is certainly a function of class, and probably also varies regionally. However each illustration represents a realisation under a uniform value of ρ.

The most obvious visual effect of the spatial dependence parameter is shown in the sizes of inclusions. As ρ approaches 0.25, inclusions become larger, as illustrated by the large blobs in the ocean in the lower right illustration. The lower class membership for water around Point Arguello have produced a higher density of inclusions. When compared to Plate 17, these illustrations demonstrate the sharp difference between the two approaches to display of uncertainty: the parameters of the error model in the form of class memberships (Plate 17) and realisations of the model as a stochastic process (Plate 20).

... SUMMARY AND FUTURE DIRECTIONS

It is often argued in the GIS community that while uncertainty is endemic to spatial data and undoubtedly affects the outcomes of spatial data processing, it is best not to draw attention to it because of its complexity and potentially damaging effects on decision-making. The user 'does not want to know'. Analogous software systems, such as the statistical packages and database management systems, do not include techniques for capturing, storing and manipulating explicit information on uncertainty, so why should GIS? We believe that this argument is both intellectually unsound and disastrously short-sighted. Most spatial decisions, particularly important ones, are made in an environment of conflict and controversy. As GIS matures and becomes available to more and more parties to a debate, the naive view that the party with the GIS somehow carries greater weight will become less and less realistic, and easier and easier to attack. Pressures for better quality assurance and control are already emerging from instances of GIS-related litigation.

In the context provided by a stochastic error model, the user of an error-handling GIS has three alternatives. First, displays can omit all reference to uncertainty by showing the most likely class for each pixel. This is the most commonly encountered option, and has pervaded GIS to date, particularly in its applications to the spatial distribution of environmental parameters such as soil class or land cover class. Second, displays can inform the user about uncertainty through error descriptors, or through the parameters of error models. This option is represented here by the display of class membership probabilities. Finally, the user can be shown samples from the range of possible maps in the form of simulations under the error model. The differences between these three options are striking, and immediately suggestive. Realisations draw attention to the effects of error, and may be much more meaningful to a user who lacks a full understanding of the concepts of statistics. On the other hand, the arguments that have sustained a lack of attention to uncertainty in cartographic tradition – the desire not to confuse the process of communication, and a willingness to portray the world as simpler than it really is – presumably apply here also. We hope these illustrations, and those provided in the next chapter, will stimulate debate on these important issues. What impact can the use of these options have on the process of spatial decision-making, and where is each option most useful?

Spatial statistics is a complex and difficult field, and few GIS practitioners have more than an elementary understanding of its techniques and concepts. Moreover visual techniques are inherently convincing and communicative. Thus it seems that visualization will have to be a fundamental part of any concerted effort to handle uncertainty within GIS. Goodchild, Sun and Yang (1992) have argued that visualization is the key to user participation in the determination of the key spatial dependence parameters in spatial statistical models of uncertainty. In the case of the error model used in this chapter, visualization holds the key to involving mapping specialists in the determination of appropriate values of the spatial dependence parameter ρ.

An MPF is inherently multidimensional, and this chapter has presented a number of techniques for improving the user's ability to understand this particular form of spatial variation. However any communication system must satisfy the requirements of the user as much as it exploits the capabilities of the system, and it seems clear to us that an ideal design can only come from the experience of working with these tools in a real analytic environment.

... ACKNOWLEDGEMENT

The National Center for Geographic Information and Analysis is supported by the National Science Foundation, grant SES-10917.

... REFERENCES

Fisher P F (1991b) Modelling soil map-unit inclusions by Monte Carlo simulation. *International Journal of Geographical Information Systems* 5(2): 193–208.

Goodchild M F, Sun G, Yang S (1992) Development and test of an error model for categorical data. *International Journal of Geographical Information Systems* 6(2): 87–104

Laurini R, Thompson D (1992) *Fundamentals of Spatial Information Systems*, Academic Press, San Diego, Calif.

Leung Y, Goodchild M F, Lin Chih-Chang (1992) Visualization of fuzzy scenes and probability fields. *Proceedings, Fifth International Symposium on Spatial Data Handling*, Charleston, South Carolina, International Geographical Union, Columbus, Ohio

Mark D M, Csillag F (1989) The nature of boundaries in 'area-class' maps. *Cartographica* 26(1): 65–78

Tomlin C D (1991) *Geographic Information Systems and Cartographic Modeling*, Prentice-Hall, Englewood Cliffs, NJ

18 VISUALIZING CONTOUR INTERPOLATION ACCURACY IN DIGITAL ELEVATION MODELS

J. WOOD

... INTRODUCTION

Digital Elevation Models (DEMs) often form an important component of spatial databases and many GIS operations, such as slope and aspect calculation, are specifically designed to manipulate them. When combined in applications, these operations can be highly sensitive to quality of the elevation data used. Applications that can be highly dependent on DEM data quality range from drainage feature identification (Lee et al., 1992) and geomorphological characterisation (Dikau, 1989), to viewshed analysis (Fisher, 1991a, 1992) and radio wave propagation (Kidner et al., 1990). One frequently used method to derive DEMs is to interpolate the isometric lines (contours) that define a surface (MacEachren and Davidson, 1987). This chapter will examine the effect on data quality of this interpolation process using visualization as its principal analytical strategy.

As mentioned in the introduction to this section, the informal use of the term 'error' actually encompasses a number of different concepts. Ultimately, the usefulness of a DEM is determined, at least in part, by the difference between elevation values defined by the DEM and the 'true' values of the surface it is modelling. This discrepancy can be considered to be 'error'. However, this study, in common with many other GIS applications, considers the effects produced by the conversion of one surface model (contours) to another (DEM). This discrepancy can be described by the term 'accuracy'. There is a further term 'data quality' which recognises that the application and purpose of the data being used also determine its effectiveness (Beard et al., 1991). Thus, an element of indeterminacy is introduced if a study DEM 'error' does not consider the context in which the DEM is to be used. Visualization provides a mechanism which can account for this indeterminacy.

Those who use GIS are not always aware of the issues surrounding data quality, accuracy and error. Partly, this is due to a lack of tools available for dealing with such issues and users will often lack the knowledge of the types of questions to ask (Beard et al., 1991). Even with this knowledge, there is a lack of research methods available to accommodate it in subsequent spatial analyses (Goodchild and Gopal, 1989). The second purpose of this chapter, therefore, is to suggest how visualization using existing GIS functionality may be used to make users at least aware of some of these issues.

... QUANTIFYING DEM ACCURACY

Different types of potential DEM error have been identified by Carter (1989), and several studies have shown how we can measure the propagation of error through the application of GIS operations on erroneous data (e.g. Walsh et al., 1987; Chrisman, 1989; Burrough and Heuvelink, 1992). There have also been attempts to build true stochastic error models as in the previous chapter (see also Goodchild et al., 1992). However, to model error and its propagation, some information on the accuracy of the data being used must be sought.

The Ordnance Survey of Great Britain provide limited information on the accuracy of their digital data. This is usually in the form of a general statement that does not appear to be specific to any particular data. They suggest that the Root Mean Square Error (RMSE) of their 1:50 000 contour data varies from 2m in hilly rural areas to 3m in urban lowland areas. They also suggest that the interpolation of these contour data only results in a minimal degradation in accuracy of 0.1–0.2m RMSE (Ordnance Survey, 1992).

The United States Geological Survey provide a little more useful information by classifying their DEMs into three levels of quality each with their own RMSE limits (USGS, 1987). However, these values are global, and provide no information on how accuracy varies across the DEM. Several studies have attempted to overcome this by relating accuracy to measurable surface quantities such as slope (MacEachren and Davidson, 1987). What is clear is that any useful study of DEM accuracy must consider the relative accuracy across the surface.

... CONTOUR INTERPOLATION METHODS FOR PRODUCING DEMs

Many methods exist for interpolating linear data to form continuous surfaces, and have been reviewed elsewhere (e.g. Burrough, 1986). The most appropriate method will depend on the nature of the contour data as well as the purpose and application of the derived DEM. This chapter will consider four methods of interpolation. They have been chosen, not because they are necessarily most appropriate for the task, but because they are readily available, or are known to produce particular interpolation artifacts, or reveal something of the relative variation in interpolation accuracy.

CONTOUR FLOOD FILLING

For each DEM cell to be interpolated, this process searches for the nearest neighbouring contour in all directions. The path of steepest slope passing through the central cell is then calculated. A linear interpolation down this line is used to calculate the elevation at the position of the centre of the DEM cell

(CERL, 1991). This routine is available in the public domain GIS GRASS as *r.surf.contour*.

INVERSE DISTANCE WEIGHTING

For each cell, this process searches for the *n* nearest points on neighbouring contours, where *n* is an integer set by the user. The weighted average of all *n* points is calculated where the weighting is an inverse squared function of the distance of each point from the centre of the interpolated cell. This is one of the commonest and simplest interpolation methods and is available in GRASS as *r.surf.idw*. This method of interpolation is best suited for data that are relatively uniformly distributed (ie. not contour data).

SIMULTANEOUS OVER-RELAXATION

Local patches of surface can be mathematically modelled given the inputs of sampled elevation values. A simplifying assumption can be made that the interpolated surface satisfies Laplace's equation. Simultaneous over-relaxation is the iterative technique used to solve this partial differential equation (Press et al., 1988) to produce the continuous surface. The GRASS routine *r.surf.sor* is used to interpolate using this method.

LINEAR SPLINE FITTING

A fourth interpolator has been coded for this study with the express purpose of producing a map of the relative accuracy of the interpolation process. It is based on the method suggested by Yoeli (1986). The algorithm is as follows. For each cell to be interpolated, search outwards for the nearest two contour neighbours in eight equally spaced directions (N, NE, E, SE, etc.). This produces four profiles (N–S, NE–SW, etc.) each of which contains four points. For each profile, a cubic spline function is fit around these points, allowing a value at the centre of the grid cell to be interpolated. The interpolation can be constrained to fall within one contour interval of the neighbouring contours. To impose this constraint is to assume that any values outside this range would be represented by contours themselves. In effect, this interpolator produces four DEMs, one for each profile direction. Each cell in the final interpolated DEM is then calculated by taking the weighted average of the four profile DEM cells where the weighting is inversely proportional to the square of the distance between adjacent contours in that direction.

The relative accuracy of the final interpolated DEM can be described by the variability in elevation at each cell in the four profile DEMs. A high degree of accuracy is preserved when all four profile DEMs have similar values. The greater the discrepancy between interpolated cells in different directions, the greater the loss of accuracy. Thus the distribution of elevation values for each cell around its weighted average, allows the RMSE to be calculated.

...GIS BASED VISUALIZATION PROCESSES

Visualization is not limited to the use of visualization packages, nor should it be restricted simply to rendering images. Visualization can be regarded as a descriptive and analytical methodology. The visualization process does not necessarily start with the display of data, but includes the preceding steps used to derive those data. Consequently, the functionality of existing GIS can be used for the entire visualization process. All of the images used in this chapter (as well as the cover of this book) were produced using standard GIS functionality of the GRASS GIS.

To illustrate some of the visualization techniques available to the GIS user, the four interpolators described above were used to interpolate from Ordnance Survey 1:10 000 digital contour data shown in Figure 18.1. The study area is

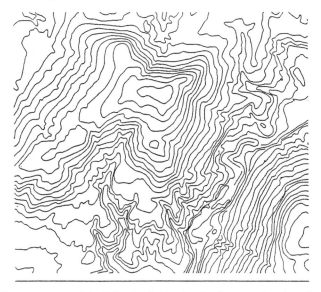

Figure 18.1 Contour model of the Peak District Study area showing the Ordnance Survey digital contour data used for interpolation, with contours at 10m intervals

located on the southwest margins of the English Peak District and covers approximately 12km² centred around the UK National Grid Reference (397200, 392500). The topography is rolling moorland overlaying Millstone Grit rocks and has elevations ranging from 70m to 300m. In all cases DEMs were interpolated to a horizontal resolution of 10m and a vertical resolution of 1m.

SIMPLE RENDERING TECHNIQUES

2D PROJECTIONS

Most GIS are equipped with several simple methods for rendering raster images on the screen. One of the commonest is to represent each cell in the raster with a colour, and display the resultant two-dimensional coloured image.

When applied to a DEM, this has the effect of producing concentric coloured bands that suggest contour lines along colour boundaries as shown in Plate 21. Images produced in this way give a broad indication of topography, and may indicate any gross errors in the data. Although one of the commonest methods of displaying data, being close to the traditional cartographic representation of elevation, this method is not very discriminating. The relatively small number of coloured bands in comparison with the elevation range means that significant changes in elevation have little effect on the image. Although minor differences can be observed between the two images in Plate 21, it is not clear whether one is more accurate than the other.

3D PROJECTIONS

Many GIS have the ability to plot surfaces on some three-dimensional projection. In its simplest form, two sets of interlocking orthogonal lines trace out elevation, or some other quantitative attribute, as in Figure 18.2. Known as a fishnet or block diagram, this form of rendering is closer to what we might regard as a 'real view' of topography. It does not suffer from the quantisation of elevation into bands, allowing more subtle errors to be observed. However, the information conveyed by such images is very dependent on various arbitrary selections of viewing parameters such as vertical exaggeration, line frequency and viewing direction. Three-dimensional projections can be enhanced by overlaying additional images on a three-dimensional surface projection. This 'drape' may take the form of the same elevation information, or additional but related information such as slope. This technique is effective in conveying two, possibly independent, spatial variables without overloading the image with too much information. Overlaying images with a high degree of realism such as shaded relief, as in Plate 22, or remotely sensed colour composites, can exploit our tendency to 'make sense' of visual images. Errors that deviate from what we expect an image to reveal can become more obvious.

FREQUENCY HISTOGRAM

While retaining the objective of identifying the spatial distribution of accuracy, it can be informative to examine visually some of the aspatial statistical properties of elevation distribution. One of the simplest statistical representations can be visualized by plotting the frequency distribution of elevation as shown in Figure 18.3. The interpolation routine used results in adjacent peaks and troughs in elevation. Because the elevation difference between adjacent peaks is equal to that of the original contour data, it can be inferred that the pattern is an artifact of the interpolation process.

PRE-RENDERING TECHNIQUES

SHADED RELIEF MAPS

Producing images with a degree of realism can be useful in the detection of error, since many sources of accuracy loss produce 'unrealistic' artifacts. Realistic images also require less additional training of the viewer in their interpretation. Many GIS have the ability to produce some type of shaded relief map from a DEM. This can be done in a variety of ways (e.g. Yoeli, 1967; Brassel, 1974; McLaren and Kennie, 1989), although the simpler algorithms that use

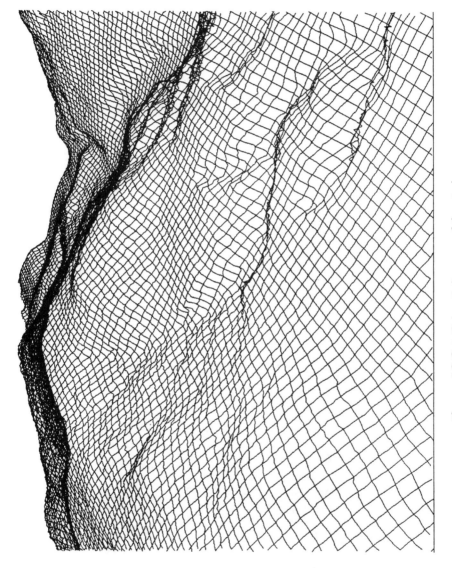

Figure 18.2 A 'fishnet' diagram of the study area

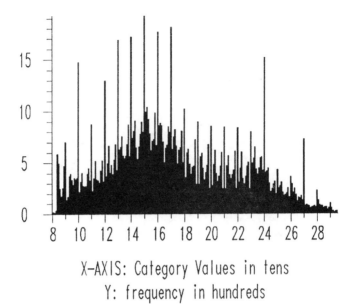

X–AXIS: Category Values in tens
Y: frequency in hundreds

Figure 18.3 Frequency histogram of elevation

only local neighbourhood information are the most useful for error detection, since it is local variation that is most prone to error. When interpreting images such as shaded relief maps that have an arbitrary lighting direction, it is possible to give a selective impression of the features in the image. This is especially true for linear features, and should be borne in mind when interpreting them. The shaded relief map of the surface produced by the inverse distance weighted interpolator showing clear terracing is shown as Figure 18.4. The breaks in

Figure 18.4 A shaded relief map

174

slope between terraces coincide with the midpoints between adjacent contour lines, and therefore reveal something of the interpolation process. The *n* nearest points selected by the algorithm for every interpolated cell will usually lie along the same contour line. Consequently most cells will retain the value of their nearest contour height. Only at the midpoints between contours will the nearest *n* neighbours switch from one contour to another.

SLOPE AND ASPECT

Slope and slope direction are commonly calculated by GIS. There are many ways in which these measures may be calculated (see Skidmore, 1989 for review), most of them use a local 3 by 3 or 5 by 5 neighbourhood. As with shaded relief maps, these localised measures can be useful for visualizing error, but caution should be exercised when interpreting slope and aspect maps because of their sensitivity to error. The slope map of the surface produced by the simultaneous over-relaxation interpolator is shown in Figure 18.5. Some evidence of terracing is visible in this image although it is not as prominent as that produced by the inverse distance weighted interpolator.

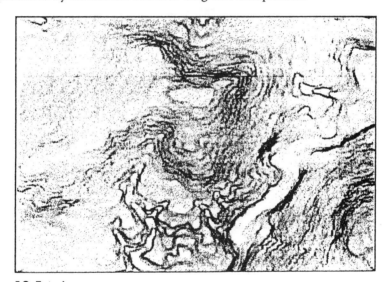

Figure 18.5 A slope map

The slope direction of the surface produced by contour flood filling is displayed using a two variable colour scheme in Plate 23. An aspect key is provided by the inset which shows an exact hemisphere shaded using the same scheme. That it appears to be a sharply pointed cone is exactly what would be expected using the aspect variable alone. The map itself gives a more realistic 'hill shaded' type of image while differentiating between slopes in all directions. This image reveals flat-topped peaks in the centre and towards the north of the study region. These coincide with the highest contour lines that represent peaks and are clearly artifacts of the interpolation process as they are not evident on surfaces produced by other methods. The same image also reveals horizontal and vertical striations around the outer boundaries of the surface. These are produced by poor extrapolation beyond the outermost contour lines.

Both features might be expected by GIS users when interpolating contour data, but by visualizing the results of interpolation, the full impact of the effects is made clearer.

CONVOLUTION FILTERING

The process of convolution filtering is common in image processing and involves creating an enhanced image from the manipulation of the local neighbourhoods of a source image (see for example, Schalkoff, 1989). This can be a useful process as error tends to dominate the local distribution of elevation cells. Most raster based GIS have the ability to perform convolution filtering, either using pre-defined or user-defined filters. Since they emphasise high frequency spatial variation, edge enhancement filters are most useful in error detection. Filters of this type may be modified to identify orientation structures in the DEM. Figure 18.6 shows the results of applying the Laplacian edge

Figure 18.6 Results of applying a 3 by 3 Laplacian filter to the DEM

enhancement filter to the DEM interpolated by linear spline fitting. Unlike some of the previous images, there is no arbitrary lighting direction, and therefore features that are visible in this image are a product of the interpolation rather than just the visualization. There is some evidence of the original contour distribution, particularly in the northwest and southeast of the image. There is also evidence of linear striations that correspond to the four profile directions used for interpolation.

QUADRATIC APPROXIMATION

Just as slope and aspect can be calculated from polynomial approximations of local neighbourhoods, so can further measures. Evans (1979, 1980) fits an approximation of a quadratic function to local 3 by 3 neighbourhoods from which, in addition to slope and aspect, profile convexity (second derivative of elevation) and plan convexity can be calculated. These measures effectively identify breaks in slope and aspect. They can be calculated in GIS that have either the ability to use map algebra, or flexible local filtering options (for details see Evans, 1979). Figure 18.7 shows a map of profile convexity for the surface produced by simultaneous over-relaxation. Even with this interpolator, there is evidence of the original contour distribution.

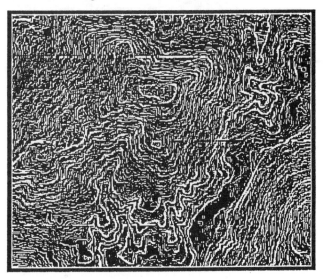

Figure 18.7 Local profile convexity

VISUALIZING ROOT MEAN SQUARE ERRORS

One of the reasons for using linear spline interpolation is to allow the calculation of a root mean square error value for every cell in the interpolated DEM. A more explicit indication of accuracy loss is afforded by visualizing this RMSE distribution as an 'accuracy map'. Figure 18.8 shows the accuracy map produced by the linear spline fitting interpolator. The absolute range of RMSE values is not as important as the relative variation in accuracy. Clearly the lineations that correspond to the profile directions used in interpolation cannot be generalised beyond this particular method. Accuracy decreases away from contour lines, especially towards the edge and corners of the image. Areas of greater accuracy loss are evident where contours diverge (for example, just northeast of the centre of the image). Since these diverging contours describe measurable features of the DEM, it would be possible to identify those areas of potential error without calculating RMSE values.

... CONCLUSIONS

Visualization is an effective tool in analysing DEM accuracy. It is suitable both because of the flexibility in interpretation of images and as a method of easily conveying the significance of error to GIS users. It is possible to use effective visualization techniques with most existing GIS without recourse to more

177

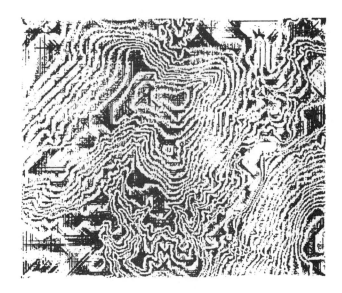

Figure 18.8 Local Root Mean Square Error (RMSE)

sophisticated visualization packages. This is important if we wish to encourage both visualization and an analytical and descriptive methodology, as well as a greater understanding of error in spatial databases.

Although visualization is potentially an analytical methodology, it is not a substitute for more conventional numerical analytical tools. It can be seen as a complementary set of tools to be drawn upon where necessary. The communication of data uncertainty issues to GIS users has been shown as a useful application of these tools.

Note: All interpolations are applied to the Ordnance Survey's 1:10 000 scale Digital Contour Data, Crown ©.

...APPENDIX TO CHAPTER 18: PRODUCING THE COLOUR PLATES

All the images and analyses were produced using the public domain Geographical Resources Analysis Support System (GRASS) developed by the US Army Construction Engineering Research Laboratory (USA CERL). GRASS is a raster based system with vector capabilities designed for UNIX workstations. Further details can be obtained from Academic Support for Spatial Information Systems, Midlands Regional Research Laboratory, University of Leicester, LE1 7RH, UK or from GRASS Information Centre, PO Box 3879, Champaign, Illinois 61826–3879, USA.

The display used for the cover of this book and the two images shown as Plate 22 (a) and (b) were produced using commonly available GIS functions. Each consists of three parts. The 'sky' was produced using a distance buffer from the top row of the raster. This gave a continuous raster surface that was rescaled to range from 100 along the top row of the image to 199 on the bottom row. The 'stars' were produced by overlaying a further layer that contained 0.1 per cent randomly allocated points assigned a value of 200. Finally, these numbers were assigned colours ranging from 100 as black to 199 as blue and 200 as yellow. The 'moon'/'sun' hemisphere was created by producing a distance buffer similar to that used for the sky. A second layer was created with values ranging east–west rather than north–south. Both layers were rescaled to range from 1000 so that 0 was at the centre of the image. These layers were called X and Y. A third, Z layer was produced by combining X and Y as follows:

$$Z = SQRT(R*R - (X*X + Y*Y))$$

where R is the radius of the sphere required.

This operation can be achieved in a GIS that has map algebra or flexible overlay facilities but for systems that store rasters as integers some degree of anti-aliasing is required. This can be done simply by adding a layer of small random perturbations and then using a smoothing filter. The final step was to mask the layer so that only positive values of the sphere were included.

Shaded relief (Plate 22 (a)), aspect (Plate 22 (b)) and slope (cover) were calculated for both the elevation surface and the hemisphere. A user-defined colour-table was used for slope and aspect. Each of these new layers was then draped over the elevation surface which itself was rendered over the sky and hemisphere to produce the composite image.

... REFERENCES

Beard M K, Buttenfield B P, Clapham S B (1991) *Visualizing the Quality of Spatial Information*, National Center for Geographic Information and Analysis Techinical Report 91–26. Santa Barbara Calif.

Brassel K (1974) A model for automatic hill shading. *The American Cartographer* 1(1): 15–27

Burrough P A (1986) *Principles of Geographical Information Systems for Land Resources Assessment*, OUP, Oxford (see especially Chapter 8: Methods of interpolation, pp. 147–166)

Burrough P A, Heuvelink G B M (1992) The sensitivity of Boolean and continuous (fuzzy) logical modelling to uncertain data. *Proceedings, EGIS 92, Munich* 2: 1032–1041

Carter J R (1989) Relative errors identified in USGS DEMs. *Proceedings, Auto Carto 9*, American Congress on Surveying and mapping, Bethesda, Maryland, pp. 255–265

CERL (1991) *The GRASS 4.0 Reference Manual*. US Army Construction Engineering Research Laboratory, Champaign, Illinois

Chrisman N R (1989) Modeling error in overlaid categorical maps. In: Goodchild M F, Gopal S (eds) *The Accuracy of Spatial Databases*, pp. 21–34

Dikau R (1989) The application of a digital relief model to landform analysis in geomorphology. In: Raper J (ed.) *Three-Dimensional Applications in GIS*, Taylor and Francis, London, pp. 51–77

Evans I S (1979) An integrated system of terrain analysis and slope mapping. Final report on grant DA-ERO-591-73-G0040, University of Durham, UK

Evans I S (1980) An integrated system of terrain analysis and slope mapping. *Zeitschrift fur Geomorphologie*, Suppl-Bd 36: 274–295

Fisher P F (1991a) First experiments in viewshed uncertainty: the accuracy of the viewshed area. *Photogrammetric Engineering and Remote Sensing* 57(10): 1321–1327

Fisher P F (1992) First experiments in viewshed uncertainty: simulating fuzzy viewsheds. *Photogrammetric Engineering and Remote Sensing* 58(4): 345–352

Goodchild M F, Gopal S (1989) (eds) *The Accuracy of Spatial Databases*, Taylor and Francis, London

Goodchild M F, Sun, G, Yang S (1992) Development and test of an error model for categorical data. *International Journal of Geographical Information Systems* 6(2): 87–104

Kidner D B, Jones C B, Knight D G, Smith D H (1990) Digital terrain models for radio path profiles. *Proceedings of the 4th International Symposium on Spatial Data Handling*, Zurich, 240 pp.

Lee J, Snyder P K, Fisher P F (1992) Modelling the effect of data errors on feature extraction from digital elevation models. *Photogrammetric Engineering and Remote Sensing* 58(10): 1461–1467

MacEachren A M, Davidson J V (1987) Sampling and isometric mapping of continuous geographic surfaces. *The American Cartographer* 14(4): 299–320

McLaren R A, Kennie T J M (1989) Visualization of digital terrain models: techniques and applications. In: Raper J (ed.) *Three-dimensional applications in GIS*, Taylor and Francis, London, pp. 80–98

Ordnance Survey (1992) *1:50 000 Scale Height Data User Manual*, Ordnance Survey, Southampton UK

Press W H, Flannery B P, Teukolsky S A, Vetterling W T (1988) *Numerical Recipes In C – The Art of Scientific Computing*, Cambridge University Press, Cambridge

Schalkoff R J (1989) *Digital Image Processing and Computer Vision*, Wiley, New York

Skidmore A K (1989) A comparison of techniques for calculating gradient and aspect from a gridded digital elevation model. *International Journal of Geographical Information Systems* 3(4): 323–334

US Geological Survey (1987) *Digital Elevation Models – Data Users Guide*, US Geological Survey, Reston Virginia

Walsh S J, Lightfoot D R, Butler D R (1987) Recognition and assessment of error in geographic information systems. *Photogrammetric Engineering and Remote Sensing* 53: 1423–1430

Yeoli P (1967) Mechanisation in analytical hill-shading. *Cartographic Journal* 4: 82–88

Yeoli P (1986) Computer executed production of a regular grid of height points from digital contours. *The American Cartographer* 13(3): 219–229

19 ANIMATION AND SOUND FOR THE VISUALIZATION OF UNCERTAIN SPATIAL INFORMATION

P. FISHER

... INTRODUCTION

Spatial information is fraught with uncertainty and error. A variety of conventional methods have been developed to support the paper map in the presentation of uncertainty (see Fisher, 1989, 1991c). Thus different line-types can be used to illustrate the certainty with which geological boundaries are known or inferred, and the extent to which lake shorelines and streams are intermittent. Similarly, where one of two phenomena such as soil or vegetation may be present, area symbols (colour or tone) may interdigitate (Keates, 1989). Many written reports of natural resource surveys contain considerable amounts of information on data quality. Standards can be set up for production of map series, perhaps the best known example being the National Map Accuracy Standard of the US (Thompson, 1988). For many users, even those working in technical areas, this quality information is often opaque and deeply hidden. Only the best informed soil scientists are aware of either the possible rates of error in a soil map, or how to find out any information there may be on that error. In short, except to salve the conscience of the map producers, the information on the quality of data made available to map users is almost useless. A number of visualization tools need to be developed to portray error at the same time as the original data. The increasing use of computer displays and the development of stochastic models of error present the opportunity of doing just this.

In this chapter two methods for visualizing the error are presented. In the first, error terms are integrated with the display by animation, where the locations of features are revised by random selection as the map is displayed, and it is believed that the uncertainty in the locations and attributes are communicated to the user at the same time as the information in the original map is retained. The second method uses sound to convey the error by highlighting a location and sounding either tone or rhythm in proportion to that error. The discussion considers the effectiveness of these methods. A systematic experimental testing of the effectiveness of the displays is essential but is not without problems.

... ERROR ANIMATION

RANDOMISING THE DOT DENSITY MAP

A dot density map consists of a set of polygons within each of which a certain number of objects are known, or believed, to occur. Dots are then placed at random within that polygon to portray the occurrence of the objects, at the rate of one dot for a certain number of the objects. Occasionally some areas within a polygon may be known to be under-inhabited compared with others (e.g. urban versus rural areas) and in a dasymetric fashion the dots may be arranged accordingly (Cuff and Mattson, 1982). In a traditional hard copy map this static representation may be the best way of displaying count information, and has certainly been widely used. In a computer context this need not be the case, and yet a number of computer mapping systems do produce static dot density maps. The implication of the dot density version of a population map to a user unaware of the cartographic conventions governing its creation is that wherever a dot occurs they can expect to find about that number of people. They might expect a village or some other cluster of houses to be present for which there is no basis in reality. This is a common problem with almost all maps that portray areally aggregated data.

An animated dot density map is the alternative suggested here which is believed to remove any confusion. For each polygon a list is formed of the coordinate pairs of the randomly located dots. The map is then displayed, showing polygons and dots. Dots are then selected at random, and relocated to new random positions within their respective polygon. In this way no dot display remains in a single position for long, and the removal and re-display of the dots is believed to convey the locational uncertainty of the map, specifically that the dot is only meaningful within a region, rather than at a particular point location.

RANDOMISING THE SOIL MAP

A soil map displays relatively large areas as having homogenous soil characteristics, and so its user expects to find soil type x at any location where x is mapped. This is not in fact the case, as detailed examination of USDA-SCS reports, for example, will demonstrate (Hayhurst et al., 1977). Details of included series are given and their frequencies may either be inferred from the mapping standards, or from the reports themselves (Fisher 1989, 1991b).

Fisher (1991b) presents two algorithms for taking a standard soil map and generating alternative realisations of the distribution of this noise. The first uses no information from the survey report, but simply assumes that noise occurs at a particular rate across the map and that the included soil series are those already mapped, in no particular arrangement. The second makes extensive use of the soil report information and for any mapping unit assigns cells in a raster map to a particular list of soils. Static realisations of both these algorithms are given. As with the dot maps discussed above, these static maps imply that it is known whether the soil at any particular location is the mapped soil or the inclusions.

Error animation in this display may also be achieved. A cell is chosen at random, and the simple randomised decision rule, used in the static presentation to decide whether a cell should be changed and to what soil type, is rerun. In this way the soil map is constantly changing, but after a short while the areas delineated in the soil map, and the noise, can be discriminated by a user. Interpretative soil maps can also be generated with the randomised noise. These reveal information hidden deep within the county soil report, and so enhance the usefulness of the soil information.

RANDOMISING THE CLASSIFIED DIGITAL SCENE

One of the major products from remotely sensed imagery is the thematic or classified scene in which each pixel is assigned a particular land cover type. A number of different measures of error have been developed for this imagery. Ground truth checking and the generation of the error or confusion matrix can be used to yield the overall accuracy (percentage correctly classified), and both the user and producer accuracy (see Campbell, 1987, 334–65). The confusion matrix is commonly a mere adjunct to the classified image, in paper form filed separately from the image, and in digital form completely ignored. The overall accuracy in the classified image may be used in just the same way as the first of the algorithms mentioned in the context of the soil information. Similarly, the producer and user accuracies are directly analogous to randomising specific inclusions for soil map units. The accuracy with which each pixel is classified and the possibility of a pixel belonging to other classes may also be used to determine accuracy.

The most common current method used to measure and display uncertainty in remotely sensed images is by using the likelihood of each pixel belonging to the cover type mapped. This likelihood may be used as a grey scale image behind the classified map coloured by cover type, or as z values in 2.5D display. These can be hard, even impossible, to display adequately. Animation of the uncertainty can be used by randomly assigning the pixel to any cover type in proportion to the likelihood of the pixel belonging to each cover type. Pixels can be chosen at random or systematically for evaluation, with the number of times any one pixel is displayed as a particular cover type is proportional to likelihood of the pixel being in that cover type.

...SONIC MAPS OF ERROR

SOUNDING THE ERROR

There are a number of sound variables which may be adapted to mapping error. Using currently standard hardware, it is possible to use tone, volume and rhythm to assist visualization of error (under MS-DOS volume is not usually available).

Tone can be related to the level of error. The exact implementation of tone to portray error is problematic, and the most appropriate may be a personal decision. Thus to one user, a high note may suggest an erroneous pixel, and a low note an accurate one, but to another the reverse assignment may be understood. Similarly, a number of different methods may be used to relate a particular sound to an error level. The most obvious is to use a linear relationship between tones and error levels. Thus it may seem most appropriate for error to be divided equally between the changes from one musical note to the next. Other changes may be used, however, so that there may be a linear relationship between frequency of sound (tone) and error which it represents.

Rhythm may be displayed using a single tone, interspersed with silence, so that the length of either the periods of silence or of noise can indicate the level of error. High frequency clicking can be used to indicate high error and low frequency low error, or, again, vice versa, depending upon personal preference.

Finally, it is possible to combine rhythm and tone so that the tone of the noise part of the clicking varies with error as well as the rate of the clicking.

IMPLEMENTATION

Sound can be used to 'visualize' any of the error terms discussed previously, except the dot density map. If a single value is used to summarise the error (overall accuracy), however, sound does not seem appropriate because no contrasts can be given in a single image, and so there is no reference. Application to map unit inclusions, and user/producer accuracy are also hard to conceive, but sound can be easily applied when error is known at the pixel level, as in the case of the classification accuracy of classified imagery (above), or in digital elevation models discussed in the previous chapter. Whether allowing the user to roam over the image or presenting the information to the user sequentially by automatically scanning through the image, sound can be used to report the error of each pixel when it is selected by the user and highlighted. This can communicate to the user where the image is accurate and where not.

...DISCUSSION

It is not difficult to conceive how these same approaches may be used to illustrate the error in a number of other map types in the environmental and human sciences. The resulting displays will all have in common the fact that the error is embedded in the display, and so cannot be ignored by the user of the spatial information. The main inhibition to implementing these methods on a wider scale is research on suitable parameters of error for different map types. At first sight, the computing overheads of implementing the types of displays discussed in the chapter might be thought to be prohibitive. It should be recalled, however, that the processor would otherwise be idle, and even on multitasking systems the overhead may be small compared with the benefit in portraying the uncertainty in the data.

Considerable perceptual research is necessary before displays of this type should be accepted as desirable. Questions include: do the displays achieve their declared objective of informing the user of uncertainty in the spatial data; do they assist both novice and expert users; does the speed of randomisation affect the perception of uncertainty, e.g. if 200 dots are mapped within a polygon, and the dots are then randomly relocated at some rate, does the rate of relocation affect the user's perception of the number of dots, and particularly the relative frequency between polygons; can users relate perceived sounds to error, and how many different sound variables can be used in this way?

... CONCLUSIONS

A method for using animation and sound to show error in spatial data has been presented and illustrated by three separate types of map with error terms ranging from single summary values for the whole map, to separate values per map unit, and individual values for each spatial unit (pixel). In each case, embedding the error in the display makes it impossible to ignore it, which is otherwise the tendency of the user. It is as yet, however, uncertain that map users will find the message implicit in the randomised or sonic data intelligible, and perceptual studies of this are planned. The ideas discussed in this chapter have been implemented as a set of prototype programs. Details on acquiring the demonstration software are listed in the Appendix.

... REFERENCES

Campbell J B (1987) *Introduction to Remote Sensing*, Guildford, New York

Cuff D J, Mattson M T (1982) *Thematic Maps: Their Design and Production*, Methuen, New York

Fisher P F (1989) Knowledge-based approaches to determining and correcting areas of unreliability in geographic databases. In: Goodchild M F, Gopal S (eds) *The Accuracy of Spatial Databases*, pp. 45–54

Fisher P F (1991b) Modelling soil map-unit inclusions by Monte Carlo simulation. *International Journal of Geographical Information Systems* 5(2): 193–208

Fisher P F (1991c) Spatial data sources and data problems. In: Maguire D J, Goodchild M F, Rhind D W (eds) *Geographical Information Systems: Principles and Applications*, pp. 175–189

Hayhurst E N, Milliron E L, Steiger J R (1977) *Soil Survey of Medina County, Ohio*, USDA-SCS, Washington DC

Keates J S (1989) *Cartographic Design and Production* 2nd edn, Longman, Harlow

Thompson M M (1988) *Maps for America*, 3rd edn, USGS, Reston Virginia

SECTION D

HUMAN FACTORS IN VISUALIZATION

20 INTRODUCTION: THE IMPORTANCE OF HUMAN FACTORS

C. DAVIES & D. MEDYCKYJ-SCOTT

... HUMAN FACTORS IN GIS

It is now generally appreciated within the Information Technology community that the design of an interactive system cannot succeed without considering the user of that system. Inevitably, this is also true for GIS, particularly in the context of visualization, where the ability to interact with the data is essential. Failure to consider system users results in loss of productivity when users have to focus on how to perform a task rather than on the task itself. This final section of the book discusses some of the user aspects of visualization in GIS.

In the case of GIS, there are three groups of 'users' to consider. First, there are 'hands on' users of the system. When visualizing data these users are likely to be relatively experienced in GIS use and in handling spatial information. Second, there are those who form an 'audience', viewing the results of GIS visualizations in a relatively passive capacity, for example, tactical and strategic decision makers and the general public. Last, there are the designers of the system itself. This group is often neglected when attempting a user-centred approach to system design. Yet the culture, experiences and working practices of this group significantly affect the nature of the final system.

... INTERACTING WITH DISPLAYS

Much research has investigated the design of effective visualizations of spatially referenced information using various non-computerised media such as paper maps. With exceptions, such as area and perimeter measurement, the level of interaction between such visualizations and their users is largely limited to the mental processes of detection, comparison and transformation, rather than physical manipulation of the media. Instead of physically cutting up the map or drawing on it, the user looks at the map and does the same processing mentally. Using a computer, we should expect the mental load on users to be reduced, as the system performs many manipulations of the image which they had previously performed in their own imaginations. The use of GIS visualization adds features not present when interacting with paper maps. These include:

Interaction   Users have the ability interactively to modify the parameters of
their problems at hand, getting quick results and quick modifications, in
order to encourage exploration of the data.

Movement   This takes two forms. First, as seen in Chapters 13 and 14, it is
possible to animate the data using time as a variable to display movement.
Second, it is possible for the user to interact with the system to create movement of viewer-like operations such as a change of viewpoint in DEM displays, or fly-through across DEM surfaces as seen in Chapter 18.

Multiple views   Easy production allows direct visual comparison between
views. For example, it is possible to show, in a single display, the consequence of changing class intervals on maps of socioeconomic data, or multiple displays showing different aspects of a problem.

Multiple users of the same information   Users can share ideas and processes
easily, for different purposes and in different contexts.

User-designer duality   The interactive user is able to become the graphic author,
determining the design of the visualization as well as making use of the result.

To date, there has been little research into the design of usable GIS visualization. While our knowledge about designing graphic representations is considerable, and is certainly of relevance to GIS visualization, the factors which may
make the latter unique have not yet been fully investigated. In an ideal world,
research into usable spatial visualization tools would occur in parallel with the
development of new visualization systems, but GIS visualization tools already
exist which have been designed without much knowledge on user aspects being
available. There are still gaps in our knowledge and more research is required
into the user aspects of GIS visualization. In the short term, we may be able to
apply knowledge gleaned from other pertinent areas of research. Much has
been discovered about how humans interact with information displays, in both
computerised and non-computerised environments. What then can we learn
from existing bodies of knowledge, which could help us to improve the usability of GIS visualization tools?

... COGNATE FIELDS

Information which might be useful in achieving this goal could be derived from
various existing disciplines. Below we briefly discuss some of these. The last
three are the topics of the remaining chapters in this section.

It is not immediately obvious that system design is a 'people-centred' knowledge area. However, it is difficult effectively to adapt the design process
towards a user-centred paradigm without understanding the logical, practical
and interpersonal processes involved in designing a whole IT system. Specifically, for the formal incorporation into the design process of knowledge about
human behaviour, the formal methods adopted by design teams must themselves be altered where necessary. In the area of human factors, much has been
written about this necessity, and in this respect GIS designers face the same
challenge as designers of any IT system.

Graphic representations contain semantic information that communicates a
purpose or graphical message. Thus, by translating data and relationships

between data, including non-visual data, into graphic representations, users can perceive and reason about spatial relationships in a direct way. Some graphic representations display some types of information better than others. Research into effective spatial information presentation, particularly maps, has been going on for many years and is now substantial (Bertin, 1967; Robinson, Sale and Morrison, 1978; Tufte, 1983). Much is also known about graphic design in general including issues such as symbol design and perception, the use of colour and type, and so on. Many of the principles and guidelines that have been derived are relevant to computer-based visualization, but caution is needed, because this knowledge has not been empirically proven to be applicable in the computer environment. As cartographers have been aware for many years, the principles of good graphical design frequently conflict with one another. The skill of the cartographer is in successfully balancing these conflicts. While more specific investigations into the display of information generated by computers are happening, our knowledge of design principles for visualization is still limited. The result is that there are few GIS which incorporate visualizations based on sound graphic design. We need to understand what is meant by 'optimal' in the design of the graphic itself, the combination of graphics and the way in which the user interacts with them and makes decisions based on the information understood from them.

Displays of data in GIS typically present data visually on a computer screen, or on paper, in the forms of charts, tables, maps, text or pseudo three-dimensional images. Other sensory modes can also be used to represent data including hearing, tasting, smelling, touching and awareness of movement (kinaesthesia), but here we consider only visual representations. The process of receiving communication of information, by data visualization, requires detection of features, discrimination of content, and identification, all leading to interpretation of the data (Reed, 1973; Keates, 1982). All of these human activities have been studied by psychologists, who can provide information on how visualization might better achieve the purposes of effective communication, interpretation and understanding of data. Chapter 21 discusses some of the findings from psychology which relate to the perception and interpretation of visualizations in GIS.

The ease of interaction between the user and the tool is crucial in determining the success or failure of an interactive visualization environment. Ergonomics provides insights into human interactions with tools and the manner in which their design can contribute to improving the efficiency with which they are used. Although ergonomics provides tools and methods, such as task analysis, by which to study the use of computers, its focus is on what has already been designed and built. Introducing changes to the system, to improve interaction, at this stage can be very difficult due to technical and organisational inertia. HCI, which is one of the fields of study in ergonomics, is concerned not only with applied research but also with design practice. By its nature, the science of HCI is interdisciplinary, and has as its focus the creation of usable systems. It studies computer users, computer usage, interaction styles and techniques and tries to develop design methodologies and principles (Baecker and Buxton, 1987; Helander, 1988). Recently there has been some interest in HCI with visualization tools (Gorny and Tauber, 1990), although, as yet, this is still limited. The subject of HCI in relation to GIS and visualization is described in Chapter 22.

... PROBLEM-SOLVING, SCIENTIFIC METHOD AND EPISTEMOLOGY

Visualization is an interplay between technology and the human mind, but at the heart of this process are the ideas on people's minds. Visualization is seen to be especially important in the transfer of ideas because of the richness of visual signals and the problem-solving nature of vision. It is important therefore to understand the philosophical issues involved in visualization if we are to give proper emphasis to the role of ideas in communication, system design and graphic design. In particular, we need to begin to understand what visual knowledge is and what spatial thinking is. These issues are addressed in the final chapter of this book.

However, merely mixing knowledge and principles from these areas of research will never suffice to ensure that a system is optimised for its end users. The entire approach to design must centre on the users and their tasks. This applies not only to system builders but also to users who customise the interface and/or design actual screens of information for presentation to other users. Thus, the well-established techniques of task analysis, prototyping and usability evaluations are also appropriate to the application of GIS to successful visualization of data. Above all, we have a great gulf in our knowledge of specific user-related problems with unique features of GIS visualizations. This means that many of the issues raised in this section of the book can at present only be raised and not definitively answered. These issues present a challenge for current and future research.

... REFERENCES

Baecker R M, Buxton W A S (1987) (eds) *Readings in Human Computer Interaction: A Multi-Disciplinary Approach*, Morgan Kaufman, Los Altos, Calif.

Bertin J (1967) *Semiologie Graphique. Les Diagrammes – Les Reseaux – Les Cartes*, Mouton-Gauthier-Villars, Paris-La Haye

Gorny P, Tauber M J (1990) (eds) *Visualization in Human Computer Interaction: 7th Interdisciplinary Workshop on Informatics and Psychology*, Springer-Verlag, New York

Helander M (1988) (ed.) *Handbook of Human–Computer Interaction*, Elsevier-North Holland, Amsterdam

Keates J S (1982) *Understanding Maps*, Longman, Harlow

Norman D A, Draper S W (1986) (eds) *User-centred System Design: New Perspectives on Human Computer Interaction*, Lawrence Erlbaum Associates, Hillsdale, New Jersey

Reed S K (1973) *Psychological Processes in Pattern Recognition*, Academic Press, London

Robertson A, Sale R, Morrison J (1978) *Elements of Cartography*, Wiley, New York

Tufte E R (1983) *The Visual Display of Quantitative Information*, Graphic Press, Cheshire, Connecticut

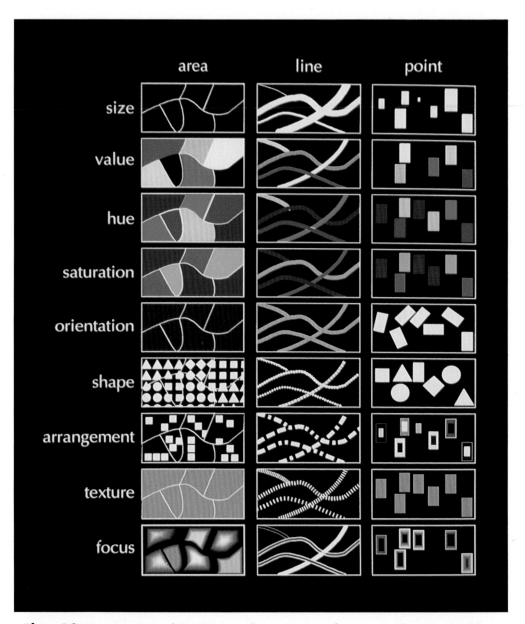

Plate 13 An extension of Bertin's graphic primitives from seven to ten variables (the variable of location is not depicted)

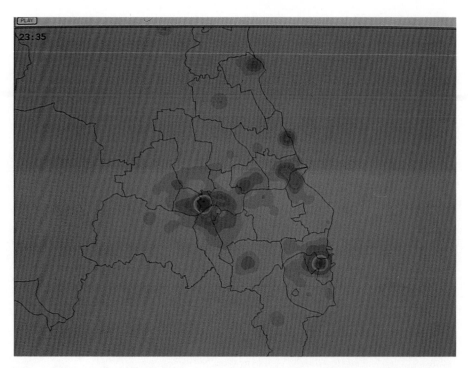

Plate 14 The expected peaks in public disorder offences after public houses and night clubs have just closed, highlighted in Newcastle city centre and other major night spots in Tyne and Wear

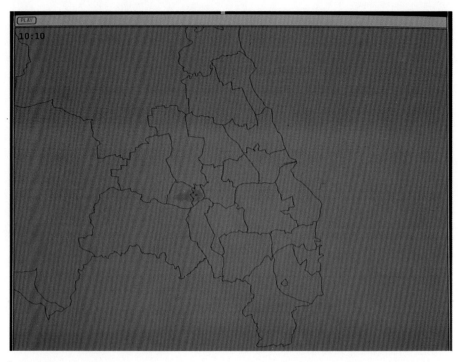

Plate 15 A typical movie frame from an early morning view of public disorders

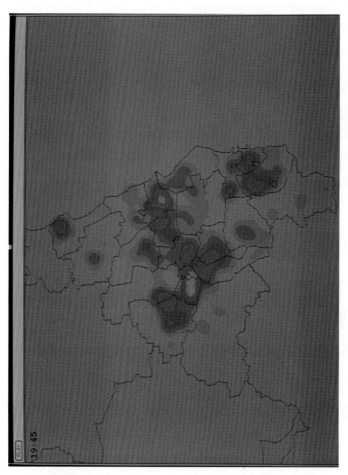

Plate 16 Reported incidences of public disorder for early evening in Tyne and Wear

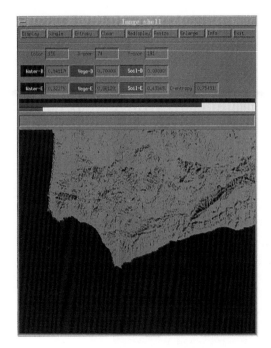

Plate 17 Fuzzy classification of part of the Santa Barbara Thematic Mapper (TM) scene, displayed by mixing red, green and blue, according to memberships, in soil, vegetation, and water classes respectively

Plate 18 Fuzzy classification displayed using grey scale images for each class: water (top left); vegetation (top right); and soil (lower left). White indicates highest membership probability

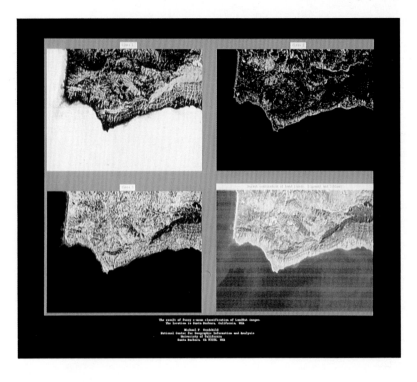

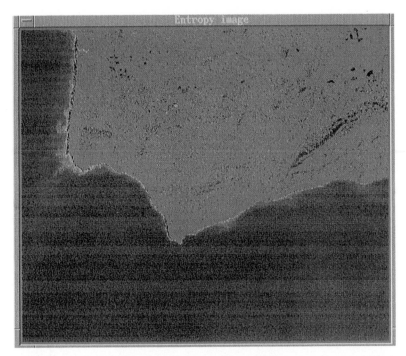

Plate 19 Entropy calculated by pixel over classes. Dark areas have values of the information statistic close to 0 (one class dominant); light areas have values close to 1 (all classes equally likely)

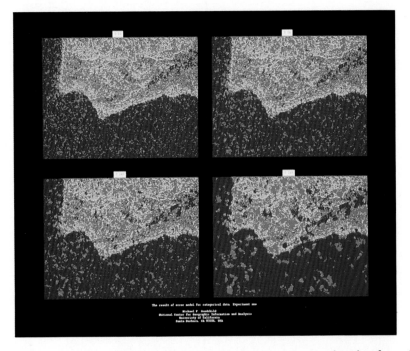

Plate 20 Realisations of the error model, for four increasing levels of spatial dependence. Note the increasing sizes of inclusions as the spatial dependence parameter increases

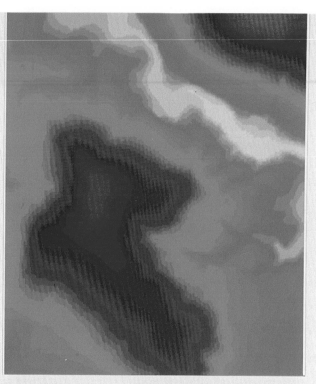

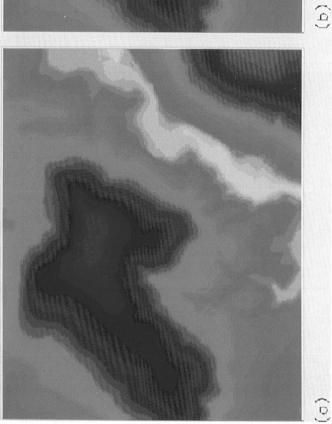

Plate 21 DEMs are commonly visualized by associating every elevation value with a colour. This indicates the broad topographic characteristics of the modelled surface but is not very discriminating. This method of visualization is used here for two forms of interpolation from the same digital contour data. (a) was produced by an inverse distance weighted interpolator and (b) by simultaneous over relaxation. This is a poor method of visualization for error detection

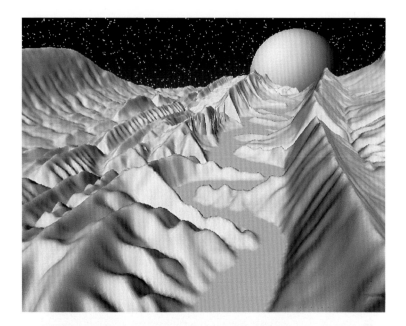

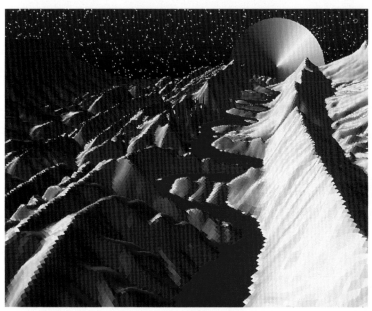

Plate 22a & b DEMs can be readily visualized using a three-dimensional surface projection over which additional information may be 'draped'. Plate 22.2(a) shows a shaded relief map draped over elevation. The view is looking southwest along River Raystown, Pennsylvania, and is approximately 5km from left to right along the edge of the DEM. The 'moon' in the background is a hemisphere over which exactly the same shaded relief has been draped. It acts as a key indicating the range of grey values and lighting direction (from the northeast). Plate 22.2(b) shows the slope aspect draped over the same surfaces. Note that the hemisphere now appears as a cone showing that aspect carries no indication of slope magnitudes, merely its direction. For details of how these images, and that used for the cover of this book, were produced, see the Appendix to Chapter 18

Plate 23 Slope aspect as an indicator of DEM accuracy. The inset shows which colour is associated with each slope direction. Visualizing DEM aspect reveals some of the artifacts of the interpolation process used to create the DEM. Edge effects and flat topped peaks can result from contour flood filling.

21 PSYCHOLOGY AND DISPLAYS IN GIS

H. HEARNSHAW

... INTRODUCTION

Displays in GIS typically present data visually on a computer screen, or on paper, in the forms of charts, tables, maps, text or pseudo three-dimensional images. Other sensory modes, such as hearing, tasting, smelling, touching and awareness of movement (kinaesthesia), can also be used to represent data, but here we consider only visual representations. The process of receiving communication of information by data visualization requires detection of features, discrimination of content, and identification, all leading to interpretation of the data (Reed, 1973; Keates, 1982). All of these human activities have been studied by psychologists, who can provide information on how visualization might better achieve the purposes of effective communication, interpretation and understanding of data. This chapter discusses some of the findings from psychology which relate to the perception and interpretation of visualizations in GIS.

... THE VISUAL SYSTEM

The human visual system comprises a receptor system, the eyes, linked to a processing system, the brain (Keates, 1982). The external visual stimulus reaches the retina, a set of sensitive cells at the back of the eye, which produces electrical signals sent to the visual cortex in the brain. The mapping between retinal cells and brain cells is complex and not fully understood, but involves both processing and selection along the way. Similar transport of received stimulation takes place in the other sensory modes. Although the perception of stimuli is complex, it has been well studied. The visual system is neither just a camera nor just a computer, but the means by which the external stimulation of light receptors is transferred into the internal long-term memory (Haber and Wilkinson, 1982).

The flow of information within the visual system is not only inwards, from external stimulus to internal storage, but outwards as our knowledge, stored in memory, is transmitted through the visual system to affect our interpretation of visual scenes and our perception and discrimination of them. For example, our knowledge that it is difficult to determine the colours of objects when they are seen against a bright, contrasting background such as sunlight (looking at apples in a tree to judge their ripeness), will suggest to us to move our point of view so that the sunlight is behind us (by walking round to the other side of the tree), so reducing the glare and contrast and enhancing our ability to perceive the colours of the objects. Knowledge influences the way we look at things. Information flows in both directions between the eye and the long-term memory.

...PERCEPTION OF VISUALIZATIONS

Stimulus detection demands certain physical attributes of the stimuli. These include desirable levels of intensity, size, duration, separation by distance, difference in form, dimension and contrast. A number of studies have produced psychophysical laws for discrimination between stimuli which give the function by which we can predict the change in response to a given change in stimulus (Stevens, 1959). The knowledge that a difference in form between two objects is easier to see than a difference in size, that contrast enhances detection, that illumination, within upper and lower limits, improves detection, and that regular, simple geometric forms are easier to identify than unfamiliar, asymmetric ones, can be applied to improve the perception of displays. A more complete discussion of stimulus attributes can be found in Held and Richards (1972). Having detected the presence of a visual stimulus, the next stage involves detection of the features in the image. To do this, we compare perceived stimuli against our learned vocabulary of features. In the visual mode, features include such things as edges, shapes, shadings, patterns, etc. In viewing a map, as opposed to another form of picture, a specialised, learned vocabulary of features is employed (Keates, 1982). The method by which features are coded and stored in memory for later recall and comparison with newly detected and coded features is described by a number of alternative theories (Reed, 1973).

Cartographers have always used this knowledge of feature detection in designing maps. Features such as contrast, size, shading, colour and form have been used carefully to communicate information in legible, unambiguous, usable maps. There are many conventions used in cartography which need to be learned by the user in order to make full use of the map. For example, the meaning of a contour line is not immediately obvious without some explanation. Thus, in order to recognise and recall the meaning of symbols, memory plays an important part in the active process of map-reading. Using symbols to represent values also leads to decisions on suitable sizes to be correctly perceived, and it is known that the shape of a symbol interacts with its size to affect the interpretation of its magnitude. Keates (1982) goes as far as to say that, because of this interaction, using a symbol to display quantitative data accurately is impossible. GIS frequently use screen map displays for data and so this question is very relevant in displaying quantitative values graphically.

Colour is an important variable by which to display information and is widely used in cartography. It is agreed that our perception of colour is a three-dimensional perceptual space (Robertson, 1988). One set of dimensions is saturation, hue and lightness, another is red, green, blue. It is reported that the ability to discriminate between saturation levels of fixed hues depends on the area of a coloured image and on its spatial separation from other coloured images. This means that if two greens of slightly different saturation are placed adjacent in a map legend, the difference between them will be easier to see than if small areas of the same two greens have to be compared when widely separated on a map (Keates, 1982). If the areas of the green patches were larger, then the colour discrimination would be easier and hence less liable to error. The implication of this for a visualization is that choosing colours to perform the required task in the display is not simple, since it depends on at least scale and separation, and has a direct effect on the interpretation of the data.

The choice is also affected by the device which will display the colours. Each display device has its own characteristic gamut of possible colours. Robertson (1988b) has attempted to clarify the choice of a hardware device for such a task by identifying its available colour gamuts. This limits the choice of display colours to ones which can be discriminated from the display device available for the task. It is not sensible to choose a colour which your screen cannot produce.

Perception of a three-dimensional scene from a two-dimensional display can easily cause misinterpretation due to ambiguities. In order to recognise depth in a real life scene we need to change our point of view and compare the two views. The relative parallax and speed of movement of items in the display which this change produces will lead us to perceive them as being different distances from the eye. Movement of the head is used to distinguish which item is the nearer (Hayashibe, 1991). This ambiguity of distance can cause major mistakes in viewing a visualization if it cannot be distinguished which of two items is the nearer. Changing the focus is also used in the real world viewing to ascertain depth but in a two-dimensional display this option is absent. Stereo images are used in GIS to display pseudo three-dimensional displays, but can easily also produce illusions, leading to errors in interpretation of the data. One source of illusion can be any spectacles worn by the viewer especially if the two eyes have been provided with different lenses. The stereo displays are unlikely to be able to compensate accurately for this.

The visual system allows us to interpret a smoothly moving image from a sequence of still ones by bridging the gaps between the images, as we do when watching a film. Our ability to fuse images together or interpolate between images is one which visualizations can use fully. One specific use of animation is to enhance our detection of change in a display. We are good at noticing the difference between two representations of data when they are displayed so that changes appear as movements. Also, perception of movement is better at the periphery of the visual field. These findings suggest that displaying changes in data by visualization would be most effective when showing changes as apparent movements at the edges of images. Animation can also be used to enhance our understanding of how the passage of time relates to changes in data. Our understanding of data on time, as a variable, is best displayed using time as the display variable. We understand the display of time in animations over time.

Perception is an active, cognitive process requiring time and depending upon memory. In looking at an image, the eyes will fix at one point and then jump to another point of fixation (saccadic movement). In the period of rest on an object they are also making very small rapid movements, or micro-saccades, as they search around the point of attention. We are generally unaware of any of these movements. The order in which saccades and fixations happen is not fully understood, but repetition is typical. The observer will return to repeat some fixations and update the stored memory of that point. It is also known that

195

some parts of an image will attract attention more than others, and that attention can be deliberately guided by the layout or nature of the display. Repeated attention is given to areas where the interpretation is ambiguous, or where there is a lot of information displayed. In creating visualizations, we should remember that not all parts of a screen display will be given equal attention.

··· INTERPRETATION OF VISUALIZATIONS

Making sense of a representation requires more than just perception of the external stimuli. Understanding a visualization is also a cognitive process. There are characteristic ways in which humans will interpret a visual representation, some of which can lead to misinterpretation, and which need to be recognised. They are now discussed.

Humans try to organise an image into a real world scene as a priority over any other interpretation, even if there are stimuli in the image which contradict this (Haber and Wilkinson, 1982; Robertson, 1991). We are skilled at doing this through regular practice in daily life. In reading handwriting, for example, letter formations rarely match the stereotypes we learned as children, but we interpret the squiggles to be appropriate letters in the context in which they appear without too much difficulty. We also try to make some organised sense of any visual image even when it appears unfamiliar and apparently meaningless (Robinson and Petchenik, 1976). In using visual images to display data to lead to understanding, it is very easy to mislead or be misled. No image is absolute, rather it is dependent on the viewer. The real world paradigm by which we try to fit every image to a representation of a real world scene is not always relevant to the world of GIS. There may be no real world visible scene relating to the data to be displayed as in, for example, a map showing a distribution of temperature. There must, however, be some essential 'abstraction' attributes of a scene however it is displayed and this is what the visualization is aiming to capture and convey.

The human will attempt to supply a complete interpretation of a partial image. The Gestalt school of psychology describes this as making the whole interpretation of an image greater than the sum of its parts. We tend not only to complete partial images, but also to provide an understanding of a complete image even when it is not there. An example of this is our understanding of an impossible staircase as a real staircase even though it can be seen that it is an image which cannot represent real life. The need for making sense overrides the accurate recognition of stimuli.

The context of a display will allow the user to select a set of candidate interpretations from the range of all possible. From this selection, a final interpretation is determined by matching the display with an image from memory. This can lead to some potential interpretations being ignored because they do not fit with the viewer's understanding of the context and, hence, their criteria for selection of candidate interpretations. Expectations of what to see can be just as powerful in determining what is seen as the stimuli themselves. Our previous knowledge affects the way we see things. In ambiguous images, such as Frisby's (1979) invertible cartoons (a picture can be seen in the drawing, but if it is inverted a completely different picture is seen), the stimuli, or parts of the image, are perceived in a way which is determined by the interpretation of the whole image. This whole image is, in turn determined by the interpretation of parts of the image. Studies of fixed or of masked images show that they fade element by element or appear in a specific order of elements. Generally the sides of features appear first, followed by angles and finally colour. Colour becomes apparent only after the whole contour becomes established. These findings seem to indicate that, in order for objects to be perceived, specific features or elements must be processed by the visual system. These features are first identified and then combined in the process of perception. However, there is a paradox here. Anything termed an element, the side of an object or an angle, could, depending on how we look at it, be called a figure. What makes an element an element is being part of a larger form. The problem is how do we know that an element is part of a larger form or pattern when in order to see the pattern we have first to construct it from the elements? We must first have an idea of a global figure in order to be able to organise local detail. Do we first organise the global view then fill in local detail or do we accumulate local detail and then create a global view, or both?

What we see in an image is not an absolute quality of the image, and may be very different from the intended, or real life, content of the visualization. This phenomenon is well illustrated in the optical illusions we all know, but is just as true in the illusions of which we are unaware. Visualization in GIS may well expose a new crop of illusions to amuse us, providing we recognise them. It may not be so amusing if a critical decision is based on a misinterpretation of data because of an optical illusion.

Memory is a necessary adjunct to the visual system in interpreting images. Displays are interpreted by comparing them with images in memory. The capacity of memory determines how many items can be stored in a feature vocabulary or can be learned as conventions, but there is a limit to a human memory. Chunking of information can improve the capacity to store it, and displaying as a picture is an efficient chunking mechanism (Haber and Wilkinson, 1982). The old adage of a picture being worth a thousand words could be usefully rewritten to say that a picture can be remembered very much more easily than the same information in a thousand words. We are very good at remembering pictures and the details they contain. The cartographer's tools of colour coding, shading and form are also effective techniques for chunking of information. Chunking can be created, for example, by colour coding in choropleth maps, or by formatting in computer program listings, where use of upper and lower case, and of brackets, can display structure within the text in a way easy to understand. In general, humans are able to recognise items from memory better than recall items, and interpretation of visualizations requires mostly recognition rather than recall, so demands on memory need not be too high. Reducing the memory load will increase the likelihood of the communication of information, via visualization, being successful.

...SPECIAL FEATURES OF SPATIAL DATA VISUALIZATIONS

The human experience of spatial relations between objects is used as a common basis of shared understanding when spatial relations are used to display structure and this is another reason why we are good at remembering images. In an image the spatial relationships between the different parts contain information about those parts. Thus, items which appear touching are seen to have something in common, and some parts can be seen to be further apart than others. These relationships are easy for us to recognise because we have lots of practice in real life. Another example of spatial relationships conveying information is in program listings where indentations use spatial relations to display structure by using the position of text to indicate groupings of its content.

GIS data displays include charts, tables and maps, all of which use the position of items in space to convey information. Maps are, of course, one of the richest sources of information display using spatial representation of data. Map tasks typically involve the need to make judgements of distance, direction and extent of objects as they are perceived.

... CONCLUSION

We have seen that interpretation of perceived visual stimuli can be affected by many things, both in the nature of the display and within the viewer's own visual system. A map will simultaneously employ many display variables, such as location, colour, symbology, size, spatial pattern and blank spaces to convey much information. A map visualization in a GIS provides much more than its equivalent paper map. Animation and movement are some of the additions already discussed in this book. Another is the duality of the role of the user as both a display designer and a display user. Visualization allows exploration of data if it is an appropriate visualization. So, the GIS user must be able both to select the right display features and interpret them correctly. In GIS, the same data are available to many users. Each user can create a different visualization of the same data. Recognition of the similarities and differences of the various displays can lead to better understanding of the data, but only if the validity of the interpretations of the displays are understood.

This chapter has discussed some of the traps for the unwary which can lead to the creation of invalid displays. It is hoped these traps can now be avoided by the reader. Good cartographers produce maps which best use the skills of humans while avoiding their weaknesses in interpreting visual displays. GIS will need to employ all these skills and add to them the awareness of how users will perceive and interpret the data communicated by visualizations. Computer screens offer new representations of data and data structure which traditional cartography cannot provide. Viewers will perceive and interpret those new representations in ways which will sometimes differ from the old ways. Using the findings of studies of human psychology can help us to make interpretation of data visualizations more reliable and effective, and hence do the job we want them to.

... REFERENCES

Frisby J P (1979) *Seeing. Illusion, Brain and Mind*, Oxford University Press, Oxford

Haber R N, Wilkinson L (1982) Perceptual Components of Computer Displays. *IEEE Computer Graphics and Applications*, May, pp. 23–35

Hayashibe K (1991) Reversals of visual depth caused by motion parallax. *Perception* 20(1): 17–28

Held R, Richards W (1972) *Perception: Mechanisms and Models*, Readings from *Scientific American*, Freeman & Co., San Francisco

Keates J S (1982) *Understanding Maps*, Longman, London

Reed S K (1973) *Psychological Processes in Pattern Recognition,* Academic Press, London

Robertson P K (1988b) Visualising color gamuts: A user interface for the effective use of perceptual color spaces in data displays. *IEEE Computer Graphics and Applications* September, pp. 50–64

Robertson P K (1991) A methodology for choosing data representations. *IEEE Computer Graphics and Applications* 11(3): 56–67

Robinson A H, Petchenik B B (1976) *The Nature of Maps*, University of Chicago Press

Stevens S S (1959) Cross-modality validation of subjective scales for loudness, vibration and electric shock. *Journal of Experimental Psychology* 57: 201–209

22 Visualization and human–computer interaction in GIS

D. Medyckyj-Scott

D. Medyckyj-Scott

... Introduction

In order for the data stored in a GIS to be of use it must be possible for us to express them in some externally useful form. To this end GIS offer the means to visualize spatial data by the generation of various types of graphical representation including 2D and 3D maps, charts, graphs, tables and images. Visualization is important at every stage of geo-processing making data accessible, manageable and comprehensible. From a task point of view visualization has applications in database management (through visual monitoring of data structure and data quality), in data retrieval (image driven retrieval), data processing (through processing) and data representation (dynamic mapping, solid modelling) (ten Velden and van Lingen, 1990). For a number of reasons, the transfer of large volumes of spatial data into graphical representations is also beneficial to human cognitive performance:

- Representations act as a memory store, freeing users to solve the problem at hand.
- They greatly assist in the rapid comprehension of complex spatial relationships, sometimes ones which are unexpected or absent in the original data. An example is the closeness of isobars on a map as a direct indication of the wind speed.
- Because the comprehension of the data is easier, performance on a task involving use of those data will tend to improve.
- Users are better able to understand the problem when the data are presented graphically. They are therefore more likely to make better, more rapid decisions.
- Users are more motivated and comfortable, and therefore in a better mental state to execute tasks.

The idea of considering GIS as a scientific visualization tool is relatively new but, to a certain extent, the visualization capabilities of GIS existed in the earliest systems through map generation and display. Even so, viewing GIS as a visualization tool offers the prospect of exciting additions to the functionality available. This ability to cope with large amounts of spatial data will be of little use, however, if users find systems too difficult to use. Appropriate representations that convey relevant information as objectively and effectively as possible are obviously required but so also are the means to interact with and control the representations.

... WHY INTERACTION IS IMPORTANT TO VISUALIZATION

Interaction is fundamental to spatial visualization. The process by which a user explores, correlates and comprehends spatial data is by its nature interactive and iterative. Users benefit from the ability to modify interactively the parameters of their problem and to observe the effects in real time. This requires both interactive graphics environments which allow the user to focus on the problem at hand, interfering as little as possible with the user's explorations, and flexibility in manipulation as well as representation of the data. The ease of interaction between the user and the GIS is therefore crucial in determining the success or failure of an interactive visualization environment.

... HUMAN–COMPUTER INTERACTION (HCI)

Human–computer interaction is the process by which the user and the computer communicate with each other to perform work effectively. User–computer interaction can be thought of as a sequence of events with each event comprising an execution phase and evaluation phase, the latter acting as a feedback for the former. If a GIS is to support interaction then it must support both the execution and evaluation phases. Usually this support is provided through different means, for example user interface tools by which users can execute commands and the graphical and textual representations by which they can see the information known about by the GIS and the consequence of the execution of commands on this information. Norman (1991) refers to the idea of the gulfs of execution and evaluation: 'the mismatch between our internal goals and expectations and the availability and representation of information about the state of the world and how it might be changed' (p. 23). The gulf of execution refers to the difficulty of acting upon the information held in the GIS, the gulf of evaluation to the difficulty of assessing the state of the system, comparing what happened with what the user wanted to happen.

Bridging the gulfs can be achieved in two ways. The first is by training the users so they know how to use the system. This method fails when the task they wish to perform goes beyond the initial training. The second is by designing usable systems which requires an understanding of usability. The idea of usability has been written about elsewhere (see Medyckyj-Scott et al., 1990) but generally it can be considered as a positive quality in a system that makes the interaction between user and computer easy. In evaluating the usability of a GIS we are concerned with how well the system supports user action. For example, how well does it perform in reducing the gulf of execution, and how well does it support the perception and interpretation of the state of the information in the system resulting from actions performed upon it. The aim becomes one of designing a system which bridges the gulfs, so that interactions are performed with the GIS subconsciously, and the user's concern becomes that of working with the information and not how tasks are performed.

A first step in doing this is to understand human–computer interaction. This requires understanding and solving the dynamic interacting needs of the three principal elements of any interaction, the user, the task and the tool itself, and focusing on systems in use. By studying GIS in use we can identify the modifications required to cope with changes in interaction. The latter is necessary because the use of GIS is not static and unchanging. Interaction will vary over time in the tasks performed, user competence and the ability of the system to support what is required. Through use, new interaction needs arise, either due to changes in the type of work undertaken, or as a recognition that there are problems with the current system. The next sections of this chapter will look at each of the components of interaction in turn within the context of visualization.

... THE USERS

A user is 'a person who uses an already-defined application by sitting at a workstation and interacting with it' (Hopgood et al., 1986). Users are the most variable component of the interaction equation. They are not passive receivers and users of the systems but active individuals who filter information, act selectively, organise and create new information. The knowledge which users bring to the GIS environment, their own mental and spatial concepts of the world, experience and goals are just as important in the design of usable systems as the system itself and the uses to which it is put. As a consequence, in any discussion about interaction, we must acknowledge that the end users of GIS are active individuals who will vary in many ways:

- In their ability to understand the conventions for human–computer interaction. They also vary in their preference for the way they interact with a system and the way (format, method, amount) information is displayed.
- In their familiarity with the range of functions available in the system. For example, intermittent users will rarely build up a complete model of a system and users performing relatively closed tasks may never discover some functional aspects of the system.
- In their cognitive, perceptual and psychomotor skills, for example in their problem-solving and pattern recognition capabilities, in the way they represent and think about space (Medyckyj-Scott and Blades, 1991) and in their capability for defining goals, discriminating tasks, and identifying alternatives. Execution and evaluation are most likely to be affected by differences in cognitive functioning.
- In their application and knowledge domains. The way users see and interact with a system, the concepts they use, and the meanings given to terms, will not only be determined by individual needs and understanding, but also by the cultures and practices in which they have been trained.
- In their expectations. A user's previous experience of a particular mode of interaction creates expectations about interaction in another system. Users may transfer conceptual models developed during interaction with one system to another.

All the above will vary over time, leading to alterations in abilities, comprehension, expectations and preferences. Given this kind of heterogeneity among and within users, coupled with the fact that GIS are being used for such a wide variety of purposes, there is a move towards computer systems which can adapt their own function, at least at an interaction level, to accommodate users throughout their interactions with the system (see Browne et al., 1990; Medyckyj-Scott and Blades, 1991).

... TASKS

A task is an activity which a user of a GIS needs to fulfil in order to achieve an objective. Using GIS as visualization tools offers dramatically different ways of looking at problems and achieving goals. This introduces a new set of tasks to be performed, many of which were unnecessary when the system was being used to automate old ways of working. These tasks may have radically different cognitive requirements and use radically different cognitive capacities than the tasks necessary to achieve the old goals. The difference between the cognitive requirements and capacities required to perform the original work and how it is performed with the GIS gives us some measure of the effort required by the user to learn and interact with the GIS. At the same time, a new set of tasks for the user poses new and separate problems regarding interaction. Users of GIS in this context are engaged in open-ended types of task, often of a highly complex nature, which require a lot of flexibility when interacting with the system. Unfortunately more choice can equate with more difficulty since the user has to remember a greater number of options, commands and procedures.

It follows that an obvious requirement in building interactive systems is to identify the tasks being carried out by the user when using the GIS:

- What types of task do users perform with GIS when it is specifically being used as a visualization tool?
- To what extent do these tasks vary from one occasion to another?
- How often are these different tasks performed and does this vary between user groups?

Task analysis is the generic name used to describe a variety of approaches by which the characteristics of tasks are established (see Diaper, 1989). Unless restricted to error-free performance and focused on the direct interaction between the user and the system, they are often exceedingly complex or of little use because the tasks are determined outside of any context of doing. A more useful option is to seek highly contextualised, user-centred representations of tasks (Carroll, 1990; Kellogg, 1990). Detailed scenario descriptions of how users accomplish representative tasks with the system are one example, with the decomposition of the task descriptions into parts such as input (receiving, interpreting), decision (thinking, deciding, planning) and output (communicating, manipulation), and the relationships between these parts (Shackel, 1983). Such decomposition allows us to establish the type of interaction which will take place and need to be supported. Whatever the task analysis technique used, the resulting description should capture the similarities and differences amongst tasks in their interaction requirements.

... TOOLS

A GIS can be considered as a tool designed to maintain, display and operate on spatial information in order to serve a representational function. We can conceive of several layers of representation (Norman, 1991):

- The represented world (R^W): the data model.
- The internal representation (R^I): a set of symbols which form an internal representation within the GIS of the represented world.
- The external representation (R^E): the way the GIS displays the real world which is an instantiation of a represented world based on the content of the internal representation. There may be many different instantiations of the represented world.
- The user's representation of the represented world (R^U): the mental representation (model?) held by the user.

For a GIS to be usable it must have as one of its functions the means to transform the internal representation (R^I) into a form (R^E) for display and this form must be one that is interpretable by the user, mapping onto mental representations (R^U) the user already has, or is familiar with, and triggering appropriate cognitive schemata so that the user can interpret and comprehend the representation. One of the major issues is deciding upon the mapping between the internal representation (R^I) and the external representation (R^E). The choice of external representation and the interactions permitted by the GIS affect the interaction a user can have with R^I while different external representations will allow different interactions both on R^E itself and R^I. It is possible to differentiate between two forms of external representation called control representations and display representations. They may differ in conception, physical form, location and form of interaction.

CONTROL REPRESENTATIONS

Any representation should be as natural to the user as possible so that it can be interpreted intuitively. There are varying degrees of naturalness and this can change, even for one form of representation. This explains why a few users can get used to any GIS user interface and even claim that it is easy to use. Prolonged use of the system has made the representations, the user interface, the interaction, the displays, all familiar and natural.

The form of control representation used by a GIS carries great weight in determining its functionality and utility. Given the visual environment in which the GIS user is working, it would seem that a natural control representation such as provided by graphical user interfaces (GUI's) might be appropriate. The advantage of a graphical user interface is that it capitalises on the powerful visual processing capability inherent in human perception. Components of the interface can also be made to look like the controls which users are used to dealing with in the real world, thus narrowing the gap of execution. Because they are familiar, users already know how to use them and it is therefore easier for them to interact with the underlying system. The use of metaphors in control representations is an extension of this naturalness idea. Metaphors allow users to apply some of their real-world experience to areas of a new domain in which they might otherwise have difficulty. Some metaphors are more appropriate and effective than others. For example, while the desktop metaphor has become a very common one on some computer systems, it may be inappropriate for the town planner, utility manager or ecologist (Gould, 1993). Gould and McGranaghan (1990) suggest that a geographic analyst's workbench may be an appropriate organising metaphor for GIS but this assumes users familiar with spatial analytical techniques. The design requirements for effective and efficient interaction will vary from one control representation to another. Examples include the question of whether menus should be 'deep and narrow' or 'broad and shallow', and the most appropriate dialogues. Some requirements will be generally applicable no matter what the form of control representation.

There should be fast responses to user actions. Fast visual feedback is a desirable element for visualization systems. Delays can be annoying if they interrupt the rhythm of a task or distract the user from their current line of thought. Variations in response time can be even more annoying than consistently long response times. Where long response times cannot be avoided some kind of feedback, such as a progress indicator, should be used.

The terms and concepts used should be ones with which the user is familiar. Users should be able to interact with GIS in familiar ways, thus reducing the amount of cognitive and behavioral modification required. It can be difficult to learn and use a system if different conceptual frameworks are required at different stages of use. The conceptual framework should therefore be obvious and consistent throughout the system.

It should be possible interactively to change properties of the representation. The user should be able to interactively control viewing parameters. Examples include the ability selectively to emphasise features which are pertinent to a task while suppressing less relevant information, or to modify, in close to real time, colour variables and the degree of transparency of 3D visualizations. The latter means that the user can raise the opacity of 'interesting' parts of the visualization while lowering those parts closer to the eye that may obscure areas of interest. Such facilities allow the user quickly to filter through many instances of the representation that may be uninteresting and focus on ones which are meaningful.

There should be an optimal use of screen space. Since the same data can be explored in different ways it is best that the tool can support as many representations on the screen at once as are possible and allow the user to work with these. This permits the user to compare and combine information, reducing problems of memory limitations and ordering effects. However, although multiple windows increase the perceived viewing space, they may not necessarily increase the visual scope if the user is unable to see a relationship which spans the display or is unable to locate which window holds pertinent data. Screen layout therefore becomes a major design issue. Poorly designed screens can result in difficulties in locating needed information in a representation, confusion and disorientation, information overload, and distracting information.

It should be possible for users to rearrange the screen area to present the information in a way convenient to their tasks. Norman et al. (1986) describe screen layouts based on alternative cognitive processing strategies which might be adopted by the user. For example, one common in current GIS is the 'selective attention' layout. Here the user's task requires selective attention to one window at a time and the user focuses on each in turn. Output from the GIS is sent to one window, commands are input from another and the status of the system appears in yet another. In the 'levels of processing' layout each window displays information at successive levels of analysis. Order is implied through a left to right linear array layout. Thus a location map might be located on the far left of the screen. Next to this could be a more detailed map and next to this possibly an alternative map of the same geographical area, while on the right of the screen there could be a window showing appropriate attribute data. The 'perspective' layout assists in problem-solving and decision-making. Each window displays information related to the user's current problem but from a different perspective. The user can glance from one window to another without having to retain a mental image of the previous representation.

Users should be provided with tools so that they can navigate both control and display representations. For example, the limitations of the size and resolution of the computer screen can create a problem for the user who is, in effect, looking at the visualization through a very restricted window with no information about data that is outside of the window. Users can become lost and disoriented within a visualization and unable to relate the information being presented to the real world. The provision of continuous data navigation tools which allow the user to pan, zoom and possibly rotate to an appropriate level of resolution are necessary along with supportive displays such as locational 'you-are-here' windows.

Tools should be provided for direct interaction with display representations. Examples include the following:

- A data probe which allows for local exploration and annotation. It can display the interpolated numeric value of a point in any position in the representation such as a height, the direction of flow from a point, or an arbitrary text label tied to a position in the representation.
- A local viewing tool. Pointing at part of the representation gives a 'quick look' at all the data in that area. This might be a localised colour map within a window of interest. It serves to reduce the need for a new representation to be displayed and of the user waiting for the whole representation to be redrawn. One commercially available GIS uses such a tool for displaying the local distribution of data values in a viewshed.
- A ruler tool for measuring representation distances and communicating them to the user as real world distances and positions in the presentation space.

Through the use of 'visual' programming it should be possible to create customised visualization environments. GIS tasks cannot always be fulfilled using a single function. More usually the execution of a task will involve the selection of a series of cooperative functions. This series defines the procedural task path required to achieve a particular goal. Currently GIS support for the specification of the procedural flow of tasks is limited. The tools provided are normally command language interpreters which require the user to be knowledgeable about programming structures and are thus off-putting to less computer-aware users. A similar need arose some time ago in the area of software engineering. Here, a visual language approach is now common. The advantage of such an approach is that it allows functions to be hidden from the user and permits an implementation of tasks which maps onto the user's conceptual model of GIS use. Campari et al. (1990) have built a system which uses software building blocks presented to the user as icons which can be used interactively to implement application specific subtasks as procedural flow diagrams. Draper and Waite (1991) also use the data-flow metaphor in their *Iconographer*, a system which allows representations of discrete data to be specified by the use of visual programming techniques. The tools provided by the system allow users to develop visualizations quickly by connecting customisable components. A few commercial visualization systems now provide similar interaction tools (see Chapter 6).

DISPLAY REPRESENTATIONS

Much has been said about display representations in the rest of this book, so only a few points will be made here. First, the idea of natural mappings is important. A simple example of natural display mapping is the effectiveness of relief shading on 2D maps over the contour map for visualizing landscape (Phillips et al., 1975). In choropleth maps an additive ordered sequence of shading densities (darker equates with more) is preferable to each category having a different shading (Jenks and Knos, 1961). On the other hand, a display which requires frequent referral to a legend is a sign of inappropriate representational mapping. Representation techniques based on realistic scene representations, for example surface renderings of DEMs (see Chapter 18 and Plate 22), are effective because they reduce the gulf of evaluation by being in some sense natural. The idea of naturalness can also be used in situations where the internal representation has no natural display representation. We are used to dealing with a 3D world, its structure, surfaces and coverings. As a result when we are faced with the task of finding appropriate visual representations for data variables which are non-visual, such as temperature, representing them by physical properties of a scene can be quite successful.

One of the major issues is deciding upon the mapping between the internal representation, R^I, and the external representation, R^E. Although many representation techniques exist, trying to determine which techniques to use is not easy. The task of choosing an appropriate representation is compounded if several data variables are to be portrayed. The common approach has been to rely on those representation techniques already used in the domain with which users are already familiar. We then optimise presentations to the user's perceptual and cognitive abilities. However, as the variety of visualization techniques on offer rises, guidance is required as to the correct and most effective way to present spatial information to meet a particular goal.

A first step to achieving this is in the classification of visual representations so that we can make informed decisions about which type of display representation best conveys different types of information. Some work is occurring in this area (Robertson, 1990; see Chapter 5). Two ideas are perhaps useful for identifying the effectiveness of a display technique. One is representational expressiveness, how well a representation expresses the desired information. The second is representational effectiveness. Which representation is the most effective at exploiting the capabilities of the output media and human visual system in a particular situation (Mackinlay, 1986)?

A second step is in providing the user with support in deciding which representation technique to use. This requires either good training or intelligent visualization systems which have knowledge about the expressiveness and effectiveness of different techniques based on the properties of the data, the user's objectives, design principles and presentation rules. We also need to take care that the techniques suggested to the user not only represent the data accurately but that the resulting representation provides and facilitates sound analysis, interpretation and understanding.

... WIDENING THE VIEW

Thus far in this chapter, the phrase 'human–computer interaction' has been used in a relatively narrow way. Others, such as Marsh (1990), use it more generally to describe all aspects of the relationship between the user and the system including the physical and organisational context. The latter are obviously important for the effective use of visualization tools and thus of importance when defining the HCI requirements of a visualization tool.

The GIS workplace comprises the workstation and local working environment around it and there is considerable evidence that good workplace design dramatically reduces user problems and increases productivity (Parker, 1993). The workstation can be thought of as the tool that allows the user access to the software. While not the centre of the system, it is a critical link that allows the user to interact with their data and the software. Consideration must therefore be given to the ergonomics of workstation use in visualization. Furthermore, for the application of hardware ergonomics to be truly effective it must be combined with a proper working environment (adjustable workstations, proper lighting and heating, and frequent breaks in work), ergonomic education of the users (so that they can set up well-designed workplaces), and proper working routines.

As with any IT system, it is important to understand the specific contexts of use and application or organisational domains when thinking about the design of visualization tools. What sort of problem is the GIS being used to solve – strategic, tactical or operational – and how do these vary in frequency and magnitude? What sorts of information are used? Over what time-scales do problems have to be solved? Given that organisations are continually undergoing change, the use of GIS will not be static and unchanging. New problems, coupled with the possibility of new data being captured, may make necessary new types of display and control representations. Such factors are of importance when considering the (re)design and development of all aspects of GIS including human–computer interaction. At the same time, use of the GIS may create new conditions for collective activity, new ways of coordination, control and communication and these too must be considered.

... IMPLEMENTATION OF HCI IN GIS

The general principles regarding the building of usable computer systems are that the design of the system must proceed iteratively, that it must involve early and sustained interaction with prospective users, and that it should involve empirical measurement of use in situations which are comparable to actual use (Gould and Lewis, 1985). This view puts the user at the centre of the design process. The initial task is to establish the HCI requirements of the system. As well as the usual technical requirements and the sort of data which will be held in it, this will involve an analysis of user, the tasks, workplace and organisational characteristics. At the same time, usability goals should be set against which HCI aspects, such as effectiveness, flexibility, acceptability and 'learnability' of the GIS are evaluated. On the basis of the information collected an initial design of the control and display representations may be attempted, based upon some pre-determined metaphor and a prototype built to recognised HCI standards and guidelines.

As the design process continues, there need to be methods for assessing how successful the system is at reducing the gulfs of execution and evaluation. A number of now well-established techniques exist which can be employed to test whether usability goals are being met (see Medyckyj-Scott et al., 1990). Techniques include observational studies of use of the system, performance metrics, attitudinal surveys and interviews. Once operational prototypes are available, user performance testing can take place. This involves representative target users interacting with the prototype product and draft documentation. They are asked to carry out a series of tasks which are representative of those that the system will be used to perform. Such testing can identify problems with interaction, reliability and responsiveness of display or control representations. Solutions to problems can be identified and built into the next iteration of the design.

Given the limited amount of knowledge that currently exists regarding human computer interaction in GIS as visualization tools, it is likely that initial system designs will be very experimental. Many iterations of the 'analyse, design, implement, test, and analyse again' cycle may have to occur before a successful and usable product is created.

... CONCLUSION

Given the importance of interaction to data visualization, there is surprisingly little research into how people actually interact with visualization tools. Much more work is required to gain an understanding of users, tasks and the GIS as a visualization tool. The following are just some of the things we need to discover:

- How the components of interaction influence effective interaction.
- What the interaction requirements for supporting visualization are and how GIS should be designed to satisfy these requirements.
- How effective current user–GIS interaction is when GIS are used primarily as a visualization tool.
- What the appropriate control and display representations are and how we identify their effectiveness and efficiency for given tasks.
- How GIS users interact with complex multidimensional display representations. And finally,

209

• What other tools and interaction techniques are needed for effective visualization?

... REFERENCES

Browne D, Totterdell P, Norman M (1990) *Adaptive User Interfaces*, Academic Press, London

Campari I, Scopigno R, Magnarapa C (1990) Visual language approach: a proposal for improving user friendliness and effective use of GIS. *Proceedings of the First European GIS Conference*, Amsterdam, Netherlands, pp. 1024–1034

Carroll J (1990) Task analysis: the oft missing step in the development of computer–human interfaces; its desirable nature, value and role, panel session report. In: Diaper D et al. (eds) *Human–Computer Interaction – Interact '90*, Elsevier, North-Holland, pp. 1051–1054

Diaper D (1989) *Task Analysis for Human Computer Interaction*, Ellis Horwood, Chichester UK.

Draper S, Waite D (1991) Iconographer as a visual programming system. In: Diaper D, Hammond N (eds) *People and Computers VI*, Cambridge University Press, Cambridge UK, pp. 171–185

Gould M (1993) Two views of the user interface. In: Hearnshaw H, Medyckyj-Scott D J (eds) *Human Factors in Geographical Information Systems*, Belhaven Press, London

Gould J D, Lewis C (1985) Designing for usability: key principles and what designers think. *Communications of the ACM* 28(3): 300–311

Gould M, McGranaghan M (1990) Metaphor in geographic information systems, *Proceedings Fourth International Symposium on Spatial Data Handling*, pp. 433–442

Hopgood F R A, Duce D A, Fielding E V C, Robinson K, Williams A S (1986) *Methodology of Window Management*, Springer-Verlag, Berlin

Jenks G F, Knos D S (1961) The use of shading patterns in graded series. *Annals, Association of American Geographers* 51(3): 316–334

Kellogg W A (1990) Qualitative artifact analysis. In: Diaper D et al. (eds) *Human–Computer Interaction – Interact '90*, Elsevier, North-Holland, pp. 193–198

Mackinlay J (1986) Automating the design of graphical presentations of relational information. *ACM Transactions on Graphics* 5: 110–141

Marsh S (1990) Human–computer interaction: an operational definition. *SIGCHI Bulletin* 22(1): 16–22

Medyckyj-Scott D J, Blades M (1991) Cognitive representations of space in the design and use of geographical information systems. In: Diaper D, Hammond N (eds) *People and Computers VI*, Cambridge University Press, Cambridge UK, pp. 421–433

Medyckyj-Scott D J, Walker D, Newman I, Ruggles C (1990) Usability testing in GIS and the role of rapid prototyping. *Proceedings of the First European GIS Conference*, Amsterdam Netherlands, pp. 737–746

Norman D A (1991) Cognitive artifacts. In: Carroll J (ed.) *Designing Interaction: Psychology at the Human–Computer Interface*, Cambridge Series on

Human–Computer Interaction 4, Cambridge University Press, Cambridge UK, pp. 17–38

Norman K L, Weldon L J, Shneiderman B (1986) Cognitive layout of windows and multiple screens for user interfaces. *International Journal of Man–Machine Studies* 25(2): 229–248

Parker N (1993) GIS hardware design. In: Hearnshaw H, Medyckyj-Scott D J (eds) *Human Factors in Geographical Information Systems*, Belhaven Press, London

Phillips R J, De Lucia A, Skelton N (1975) Some objective tests of the legibility of relief maps. *The Cartographic Journal* 12(1): 39–46

Robertson P K (1990) A methodology for scientific data visualization: choosing representations based on a natural scene paradigm. In: *Proceedings of the 1st IEEE Conference on Visualization – Visualization '90*, San Francisco, IEEE Computer Press, Washington, pp. 114–123

Shackel B (1983) The concept of usability. In Bennett J et al. (eds) *Visual Display Terminals: Usability Concerns and Health Issues*, Prentice-Hall, New Jersey, pp. 45–87

ten Velden H E, van Lingen M (1990) Geographical information systems and visualization. In: Scholten H, Stillwell J C (eds) *GIS for Urban and Regional Planning*, Kluwer Press, Holland, pp. 229–237

23 EPISTEMOLOGICAL ASPECTS OF VISUALIZATION

J. PETCH

... PHILOSOPHICAL BACKGROUND

This final chapter seeks first to explore the significance of visualized data for knowledge and for knowing, and second to try to clarify the role of visualization for GIS. The issue is important because the visualization of data is a process of change occurring in all disciplines. It is being accelerated by the adoption of computers for data management and modelling and as media for teaching. Understanding the role of visual data in the processes of knowing is important in two areas: first for developing visual presentation as part of strategies of research, teaching and management, and second for improving skills of analysis and argument.

As a starting point two sets of ideas are presented around which we can structure an analysis of visualization and of visual material. The first is about the roles and relations of knowledge, of the world and of people's minds. This is the concept of three worlds put forward by Popper (1975). *World1* is the world of real objects, the physical world we know and interact with. *World2* is the world of our minds, the subjective world of our mental processes. *World3* is the world of ideas, the independent world of concepts, facts and theories contained in concrete objects such as books. Popper's concept of three worlds is not a rigorous theory which is testable. It was put forward, and is used here, simply as a useful way of understanding some important ideas about knowledge and about how people interact with things and with ideas.

Of particular importance is the notion that facts, theories and ideas, as well as the physical world and people's minds, exist independently. They have a life of their own. Once expressed in some form such as books, pictures or computer files, such knowledge is in a real sense objective knowledge. It is knowledge 'without a knowing subject'. Additionally, such knowledge is richer than that intentionally created in people's minds. To illustrate this point briefly consider the system of integers 1, 2, ... n. It was created by all of us as a counting system but its richness exposed by mathematics is apparently limitless. As a system it contains much more than was put into it. So it is with all knowledge and facts. Our creations contain more than we design into them and their value and importance increase as we explore and expose their properties.

The independence of *World3* entities is perhaps best appreciated by considering the impact they have on *Worlds1* and 2. Ideas (*World3*) guide our actions

(*World1*) and our thoughts (*World2*). Consider the improbability of between twenty and thirty people meeting together at the same time and place to discuss an agreed topic after arrangements had been made months beforehand. Such things happen only because of the controlling influence of ideas, of *World1* objects by *World3*. This problem (Compton's problem; see Popper, 1975, pp. 230–32) is not what concerns us directly but its implications do.

A second set of ideas for understanding the epistemological significance of the visual is about the logic of scientific endeavour. This is an area of great controversy in science and in philosophy. The main disagreements have been about the role of the scientific community in guiding judgement, about the nature, or even existence, of any scientific method and perhaps most importantly about the status and nature of logic in science. This third issue can be resolved, though not all agree, by the demonstrable weakness of inductive reasoning and the necessary reduction of the process of the growth of knowledge to deductive logic. This reduction is expressed in many ways, as 'strong inference' by Platt (1964), as the 'method of multiple working hypotheses' by Chamberlain (1890) and as 'conjecture and refutation' by Popper (1974). What is at the core of such reasoning is the realisation that knowledge grows through a process of hypothesis or theory testing and indeed that we learn only through the exposure of the errors in our theories. The reconstructional logic of science is one of deductive reasoning. By reconstructional I mean that real scientists do not necessarily work by consciously being deductive or by always critically testing theories, but that by reconstructing situations, knowledge can be seen to have grown only through critical testing of ideas. However, many scientists do recognise the deductive nature of the logic of discovery and design experiments to test hypotheses.

The situation of deductive reasoning and critical testing is expressed elegantly by Popper's (1975) schema:

$$P1 \rightarrow TT \rightarrow EE \rightarrow P2$$

where   P1 is a problem
        TT is a tentative theory
        EE is the process of error elimination
and   P2 is the new problem.

The chain of P1 to P2 links is infinite. We experience only the newest links. The value of the schema is in expressing a number of important ideas about the growth of knowledge. It emphasises that problems derive from the previous testing of ideas, that observations are made in relation to the testing of ideas (*Worlds1* and 3 again) and that new unexpected problems arise from testing (independence and richness of *World3* things). It also emphasises that science is necessarily active. It does not rest comfortably with unsolved problems. Some problems may be ignored for a variety of reasons, but essentially, science drives itself forward through the problem recognition–testing interplay.

...PERCEPTION, COGNITION AND SPATIAL THINKING: A SUMMARY

Visual data and visually derived information are part of the complex of 'problem definition, theory formulation, testing'. The point of this chapter is to show how the properties of visual materials give them a distinct role in this complex. In order to understand how visual data work inside the complex we need to appreciate something of the process of perception and the relations of perception and cognition. These matters are receiving attention from geographers and others interested in visualization and are raised elsewhere in this book. Nevertheless, a brief summary is given here so as to lead into a consideration of the use of visual material in GIS.

The eye–brain system can best be understood not as a passive receptor but as an active, exploratory tool. Eye movement is a searching process which is directed by the brain. Information taken by the eye is selectively obtained and selectively processed and the things the brain recognises are identified by the eye but constructed and understood by the brain (see Chapter 21). Perception is thus a psychological act of interpretation. It depends on our preconceived ideas about what it is we expect to see. Our ideas direct the physical processes of seeing which are critical, goal-directed manoeuvres.

One of the questions of visual processing, of significance for knowing, is the amount of information we are able to take from visual material. Our senses seem to have severely limited capacities to transmit information. However, if the eyes are presented with complex, multidimensional images (dimension meaning physical dimensions and dimension in types of information), they are capable of transmitting large amounts of information. This is demonstrated by the ability of the mind to discriminate and to structure patterns in complex stimuli. This leads to an important point that needs to be emphasised about maps and other complex visual images, their extreme richness for conveying information. There are two aspects to this richness. The first is their multidimensional nature. The elements of maps have position, symbology, colour and relation to other elements. The second is the richness arising from the spatial representation of data. Consider two examples to illustrate the point. In the first consider a spatial representation as a line joining two dots: •——•. Consider the number of situations or environments that have been created or described by this simple graphic. There is the •, the ——, the two •s, the set of three objects, being near one or the other or far away from them all, being at an intersection of —— and •, being equidistant from the •s, and so on. The second example is of three ellipses as in a Venn diagram. Each ellipse represents a state or condition or property. How many situations of the arrangement of the three states do you imagine can be created? Depending on what you take to be valid differences there are well over forty without any consideration of nearness or sub-environments. And not only that, you can recognise any one of them at a glance and distinguish it from another.

Knowing that eye–mind work in specific ways consistent with a hypothetico-deductive strategy and that visual material can be especially rich in information is sufficient to justify a special treatment of visualization. However, there is a further issue around which we must judge the importance of visualization which is less concrete. That is whether or not there is such a thing as spatial thinking. Are there distinct elements and specific skills of spatial thinking or not? And, if so, could we consider there to be distinct methods of spatial analysis and spatial theories which are non-trivial?

Each of us can provide evidence of the way we think spatially. If you were to be asked how someone else could get from your office to your home and were to compare how different people answered this question you would see that the reaction time to recall the route and the length of explanation had little to do with the actual distances being considered. Experimental data show that such information retrieved by the mind is independent of distance. We can assume therefore that the mind does not hold map information from images simply as images or as analogues.

Experiments on recall (Evans and Pezdek, 1980) show that subjects order spatial information more along topological lines than along analogue lines, though they tend to hold analogue type information about very restricted, familiar areas. Similarly, if subjects are asked to judge distances and positions they tend to use topological data which is not always reliable. This gives rise to common errors like judging Montreal to be north of Seattle or Edinburgh to be east of Liverpool. For the limited amount of information which is held as analogues, retrieval is slow and actual areas of recall are restricted. It is not clear if this information is held as images although its recall seems to involve a scanning process. Most spatial information is held as topological and hierarchical relations. Whether either of these types is held as images in our minds is also not clear, but its retrieval is fast and coverage greater than analogue information.

Other experiments show how information is learnt from maps (Thorndyke and Stasz, 1980). It appears that acquisition of spatial information is selective or partial. It is generally oriented and only accessible in a certain way. Spatial information gained from direct experience is more orientation-free and usable. When subjects learn information from maps they use particular strategies. They mentally partition maps into subregions, they create hierarchies of regions and features, and they encode their orderly operations using visual and spatial encoding methods. In addition they employ evaluation procedures to test themselves as they do it. It is clear that some people are better at learning information from maps than others and these are more likely to employ partitioning and use encoding methods.

At a higher and more philosophical level it is possible to consider the methods of spatial reasoning as cases of the general methods of reasoning. The point has already been made that, in spatial reasoning, perception itself operates as a testing mechanism. There are, however, several forms of reasoning. John Stuart Mill outlined five which have become known as Mill's Canons and which we can use here to get a grasp on the issue. These are:

- the method of agreement
- the method of differences
- the method of agreement and differences
- the method of covariation
- the method of residuals

In spatial reasoning, using our own powers of perception and dealing with the richness of spatial data, we are able effectively to execute all of the methods more or less simultaneously and with considerable accuracy by visual inspection. We do this through examining the patterns, structures, hierarchies and so on, that we have established on the basis of our inbuilt knowledge.

...GIS, VISUALIZATION AND KNOWLEDGE

In considering the significance of GIS for knowing it is necessary to develop some appreciation of a wider set of issues relating to knowledge and knowledge representation in a machine environment. At a philosophical level, attempting to disentangle knowledge, representations of knowledge, ideas and data will lead to circular debate and semantic blind alleys. We can think about knowledge and its visual representation in non-machine and machine environments to give four main elements: non-machine knowledge; non-machine visual representation; machine representations of knowledge; and machine visualization.

Non-machine knowledge is our knowledge about the world. In formal terms it is the set of ideas, theories and laws about how things work. It can be taken to include the classes of things we recognise, their attributes and the relations between them. An example of a member of this category would be a treatise on the effects of soil hydrological conditions on tree growth with the body of hydrological, pedological, physiological and ecological ideas which formed the matrix of knowledge on which we understood how water in soil controls how a tree grows. Much of our knowledge is probably in the form of images and the processes of thinking involve images. The idea that knowledge is verbal in nature does not stand up to scrutiny but neither is it entirely in image form. What is clear however, is that visualization, the process of creating concrete images, is an extension of cerebral visualization. It is also clear that the mental processes of visualization are stimulated by visual material.

Non-machine visual representation, in our context, is the traditional visual representation of knowledge in map or graphs. Both can be seen to represent different levels of knowledge. At an abstract level it could be a graphical representation of how tree growth relates to soil water, or a nutrient flux diagram of a forest ecosystem. At a different level, visual representations contain data. A spatial example might include maps of distributions of tree types in relation to soil units described on the basis of hydrological conditions. The creation of such visual representations involves subtle and complex mental processing which is not fully understood. Chapter 2 explains the issues involved in the process of translating ideas into maps and the same sort of issues apply to other graphical representations. As was pointed out, the process of graphic design is not yet amenable to codification despite important attempts to isolate the rules (Buttenfield and McMaster, 1991).

The input of data into machines to create a machine representation is one way of codifying knowledge. The inclusion in a GIS of a map of soil water content is completely wrapped up with knowledge about trees and soil. However, such knowledge in the machine is implicit and is heavily context-dependent. It is implicit in the situation of the machine data being used for a particular purpose. Implicit knowledge probably represent the normal way in which we can say knowledge is held in machines. Until recently in GIS, this reached its highest level in the design of databases using entity-relationship analysis and similar procedures. In these there is a systematic and controlled attempt to codify knowledge. However, there is generally no rule system in a machine which will use even such structured knowledge as objective knowledge. With implicit knowledge, that process occurs outside of the machine. Another class of knowledge encoded into machines is explicit. That is knowledge which 'works' inside the machine. This is represented by certain models and by knowledge-based or expert systems. In these are various types of machine-based rules which use data as objective knowledge, derive new data from existing data, derive information and make choices. Using our tree, soil hydrology example, rules could be programmed into a knowledge-based system such that data on soil type, topography, hydrological features and so on were read from a GIS for a particular site. On the basis of the data the system could itself reach a decision about most suitable species for planting for particular purposes such as wildlife habitat, commercial forestry and ornamental. These decisions could be based on information about hydrological conditions or on inference. Creating such machine-based knowledge for such problems is a subtle process of eliciting, acquiring and representing. Much has to be drawn from the subconscious mind by questioning and discussion so that it can be made explicit and subject to processes of formalisation, refinement and structuring.

Machine visualization can be taken to be displays of GIS and related types of software. Such display involves the use of digital data and analogue processes to drive the video displays. Vector and raster data types have been used to link to drivers which have been developed to produce a wide range of visual effects. As was illustrated in Chapter 8, considerable efforts have been made to produce realistic effects, such as the display of solid objects or the illusion of flying over landscapes. Machine visualization offers possibilities beyond traditional methods which are only recently being explored. These include animation, interactive search and multimedia. Rather obviously, most of these developments, except perhaps interactive search, emulate what has happened in cinema and TV over the last fifty years or more. The difficulty, of course, is trying to make it happen using digital data on cheap computing equipment so that scientists and data managers can use the power of the digital medium in their daily work. Much of the effort of visualization in GIS has gone into reproducing graphic styles of traditional materials. The machine visualization of knowledge, in the areas of geographical knowledge, is almost exclusively a machine rendering of visual material rather than a machine visualization of knowledge. There are few equivalents of, say, the visualization of turbulent flow which, although cerebrally possible, was never achievable in traditional media. GIS display of the tree and soil hydrology problem would be video versions of the maps with appropriate graphic styles for the medium. There are probably two forces which have meant that GIS applications derive their visualization rules and methods from traditional cartography. The first is the fact that cartography is so highly developed as an empirical study or craft. The second is that the users of GIS products demand maps!

In looking at these four areas it is possible tentatively to indicate the strength of relationship between them. The two non-machine elements, representing human activity, have strong traditions of ways of expressing knowledge visually. In cartography and in visually representing models there are highly developed skills of portraying ideas and data (Tufte, 1990). GIS, as visual representations, can be seen to be influenced strongly by the non-machine (traditional map). There is very little linkage of GIS visualization to knowledge which is distinct from this cartographic linkage. In GIS most knowledge is implicit in database structures and digital map information. Only very little

217

knowledge is explicit. This strong linkage of GIS with non-machine visual representation is, I suspect, quite different from that for, say, medical imaging or astronomy. GIS will also have stronger links from knowledge to visualization in non-machine environment and between visualization in both types of environment. Its weak links are in the codifying of knowledge, especially explicit knowledge, and in the lack of directness of the link between knowledge and machine visualization.

The reason is, I think, clear. It is that geographical ideas, theories and laws are poorly developed. The codifying of spatial knowledge is extremely difficult. Paradoxically, geographical ideas, with low or weak theory content, are nevertheless strongly visual. There is no denying the considerable spatial intellectual skill in dealing with geographical phenomena. This visual content is linked to a strong tradition of visualization in cartography. It may be that there is little progress which can be made by attempting either to develop a direct knowledge–GIS link with specific visualization techniques or to attempt to codify spatial knowledge as a means of stimulating the development of GIS. If this is the case the progress of GIS visualization rests on developments elsewhere.

... VISUALIZATION: THE ISSUES

Using the concepts of *World3* objects applied to maps we can appreciate how a map or a GIS, once created, has an independent existence as a system of facts and ideas. This independent existence and the richness of the data it contains make it open to be explored fruitfully by others. They can bring their own problems and ideas to bear on the information contained in a map. The point is that the subtlety and depth of information is unequalled by any other medium.

Psycho-physical, psychological and physiological studies of vision and perception seem to support the ideas of *World3* and the schema of critical, deductive science presented above. Popper and Eccles (1977) have pointed out that the neurophysiology of the eye and the brain suggest that the process involved in physical vision is not a passive one. It consists of an active interpretation of coded inputs which is in many ways similar to problem solving. In their view our visual perception is more like a process of painting a picture than one of taking random photographs.

In other words the action of perception fits the P1 ... P2 schema of problem-solving. We can appreciate the key role of (tentative) theory (TT) by considering the problem of optical illusions or visual tricks. Our eyes seem sometimes to tell us things that are impossible and sometimes to invent things which are not there. But, of course, it is the brain which 'fills in' the illusion for itself and we easily appreciate this when we 'understand' the illusion and by visual inspection establish that it is only an illusion.

In psychological studies of visualization there is a tendency to emphasise the importance of illusions as a danger of visual presentation. In fact they are of little importance. They operate like badly written sentences which may, at first sight, cause confusion but which are easily decoded in context. In contrast, the significant psychological issue is the control of ideas on what we see.

The real problem is how can we become skilled at intellectual invention? The answer to this question can be reduced to one considered long ago by

Gilbert (1896) when he wrote on the origin of hypotheses. In brief, his answer to the question 'how can one generate hypotheses?' was to have a wide knowledge of things.

The implication for spatial reasoning is that the more we look at maps of different things and the more we deal with spatial phenomena the better we will become at recognising and interpreting them and developing the skills of spatial thinking. The progress of GIS visualization, in other words, rests principally on developing geographical ideas.

... REFERENCES

Buttenfield B P, McMaster R B (eds) (1991) *Map Generalization: Making Rules for Knowledge Representation*, Longman, Harlow

Chamberlin T C (1890) The method of multiple working hypotheses. *Science* 15: 92–96

Evans G W, Pezdek K (1980) Cognitive mapping: knowledge of mapping and real world distance and location information. *Journal of Experimental Psychology, Human Learning and Memory* 6: 3–24

Gilbert G K (1896) The origin of hypotheses, illustrated by the discussion of a topographical problem. *Science* 3: 1–13

Platt J R (1964) Strong inference. *Science* 146: 347–353

Popper K R (1974) *Conjectures and Refutations*, 5th edn, Routledge and Kegan Paul, London

Popper K R (1975) *Objective Knowledge*, Oxford University Press, Oxford

Popper K R, Eccles J (1977) *The Self and its Brain*, Hutchinson, London

Thorndyke P W, Stasz C (1980) Individual differences in procedures for knowledge acquisition from maps. *Cognitive Psychology* 12: 137–175

Tufte E R (1990) *Envisioning Information*, Graphics Press, Cheshire, Connecticut.

APPENDIX: SOFTWARE

The contributors to this volume have used a variety of software, not all of it specifically intended for either 'GIS' or 'visualization'. It is possible to recognise a continuum from the use of 'off the shelf' visualization packages, through some form of symbiosis between different software systems, to visualization conducted using totally bespoke software written by the authors and their collaborators. Only one contributor, Wood, did all his visualization using existing GIS functionality.

...VISUALIZATION SYSTEMS

A number of software packages for ViSC are now available and are reviewed in Chapter 6. They include:

AVS: Advanced Visual Systems Inc., The Causeway, Staines, Middlesex TW18 3BA, UK or 300 Fifth Avenue, Waltham, MA 02154, USA.

Data Visualizer: Wavefront Technologies, 530 East Montecito Street, Santa Barbara, CA 93103, USA.

Khoros: Department of Electrical and Computer Engineering, University of New Mexico, Albuquerque, NM 87131, USA.

Iris Explorer: Explorer is distributed with Silicon Graphics Workstations, Silicon Graphics Inc., 2011N. Shoreline Blvd., Mountain view, CA 94039–7311, USA

To create Plate 12, Dykes used Explorer software running on an SGI Indigo machine.

The graphics discussed by MacEachren in Chapter 13 were created in the Macintosh environment using MACROMIND DIRECTOR. This is a toolkit designed for multimedia presentations, desktop video production and visualization. Paint tools within the package allow creation or modification of imported images. PICTs, sounds, Scrapbooks, MacPaint, and PICS files can be imported. For visualization-animation, available tools allow interactive access to precreated text, graphics and video sequences. For animations, tools are available to do in-betweening, transformations and object movement. A graphic interface allows time to be precisely controlled and independent modules to be linked. Interactivity is achieved through a flexible scripting language (Lingo) that contains many common higher level language procedures and also allows access to external procedures. DIRECTOR is available from: MacroMind, Inc., 410 Townsend St., Suite 408, San Francisco, CA 94107; phone: (415) 442–0200.

...MIXTURES OF SYSTEMS

Plates 1 and 2 were produced by Bishop. The images were scanned using a PC-based flatbed scanner and transferred to a Macintosh computer for colour and contrast adjustment using standard software. They were then transferred to a

UNIX Workstation where Wavefront Technology's Advanced Data Visualizer software was used to convert the images to textures. The terrain models were converted from an Intergraph generated TIN model in one case and a raster DEM in macGIS in the other. Bespoke software was used to adjust the DEM in forested areas and to add edge polygons.

The work on density estimation reported by Gatrell in Chapter 9 used a variety of software. Some was 'home-grown' FORTRAN code developed at Lancaster University by Professor P. Diggle. Other work used the statistical programming language S-Plus. Research at Lancaster has added further spatial analysis functions to S-Plus, including those for density estimation. A disk of these is currently available, price £60. Please contact the author for further details. The disk is currently available only for the UNIX version of S-Plus and requires that users have an S-Plus licence. For details of how to obtain the latter contact Statistical Sciences (UK) Ltd., 52 Sandfield Road, Oxford OX3 7RJ.

The software used to create all the graphics in Chapter 11 by Dorling was the Draw package which is now resident in the ROM of all RISC OS machines currently produced by Acorn Computers. Short programs were written to create the area cartogram coordinates and the graphics files describing structures such as the Chernoff faces. The hardware was an Archimedes 410/1, used as standard in British secondary education. It is not generally available in the United States, as are few other personal computers not conforming to the IBM PC or Apple Macintosh standards and limitations. The hardware and software used has been available since 1987 from Acorn Computers Limited, Fulbourn Road, Cherry Hinton, Cambridge, CB1 4JN, UK.

... BESPOKE SOFTWARE

A surprising number of contributors wrote their own software, often to link systems or to do something new. The movie program outlined by Openshaw et al. in Chapter 14 was written in C for a UNIX platform, in this case a SUN SPARC using the SUNVIEW graphical interface.

The methods suggested and discussed in Chapter 19 have all been implemented by Fisher in Turbo Pascal for IBM PC compatibles with VGA monitors. A disk is available from the author, with both source and executable versions of the program. Interested readers should send their requests with one blank 3.5" high density disk to Peter F. Fisher, Midlands Regional Research Laboratory, Department of Geography, University of Leicester, University Road, Leicester LE1 7RH, UK.

... USING A GIS

The only explicitly geographical information systems software used was in Chapter 18 in which all the images and analyses were produced using the public domain system GRASS (Geographical Resources Analysis Support System) developed in the United States Construction Engineering Research Laboratory (USA CERL). An appendix to Wood's Chapter describes how it was used to create the cover of this volume and Plates 21 to 23. GRASS is a raster based

GIS with some vector capabilities which runs on most UNIX workstations. In the UK it can be obtained from Academic Support for Spatial Information Systems, Midlands Regional Research Laboratory, University of Leicester, Leicester LE1 7RH. In the rest of the world it can be obtained from GRASS Information Center, PO Box 3879, Champaign, Illinois 61826–3879, USA.

BIBLIOGRAPHY

Advisory Group on Computer Graphics (AGOCG) (1992) Evaluation of Visualization Systems. *Advisory Group On Computer Graphics Technical Reports*, 9

Anderson G C (1989) Images worth thousands of bits of data. *The Scientist* 3(3): 1, 16–17

Angel S, Hyman G M (1972) Transformations and geographic theory. *Geographical Analysis* 4: 350–367

Baecker R M, Buxton W A S (1987) (eds) *Readings in Human Computer Interaction: A Multi-Disciplinary Approach*, Morgan Kaufman, Los Altos, Calif.

Beard M K (1989) Use error: the neglected error component. *Proceedings AUTO-CARTO 9*, Baltimore, Maryland, pp. 808–817

Beard M K, Buttenfield B P, Clapham S (1991) *Visualizing the Quality of Spatial Information: Scientific Report of the Specialist Meeting*. National Center for Geographic Information and Analysis Technical Report 91–26. Santa Barbara, Calif.

Beniger J R (1976) Science's 'unwritten' history: the development of quantitative and statistical graphics. *Abstracts of the Annual Meeting of the American Sociological Association*, New York

Bergeron R D, Grinstein G G (1989) A reference model for visualization of multi-dimensional data. *Eurographics '89 Proceedings*, Elsevier Science Publishers BV, pp. 393–399

Bertin J (1967) *Semiologie Graphique. Les Diagrammes – Les Reseaux – Les Cartes*, Mouton-Gauthier-Villars, Paris-La Haye (see also Bertin, 1985)

Bertin J (1981) *Graphics and Graphic Information Processing*, Walter de Gruyter, Berlin

Bertin J (1985) *Graphical Semiology*, University of Wisconsin Press, Madison, Wisconsin

Bishop I D (1992) Data integration for visualization: the role of realism in public debate. *Proceedings AURISA'92, Australian Urban and Regional Information Systems Association*, Gold Coast, Australia

Bishop I D, Flaherty E (1991) Using video imagery as texture maps for model driven visual simulation. *Proceedings Resource Technology '90*, Washington, DC, pp. 58–67

Bishop I D, Hull B R (1991) Integrating technologies for visual resource management. *Journal of Environmental Management* 32: 295–312

Bithell J F (1990) An application of density estimation to geographical epidemiology. *Statistics in Medicine* 9: 691–701

Blakemore M (1984) Generalization and error in spatial databases. *Cartographica* 21(2–3): 131–139

Board C, Taylor R M (1977) Perception and maps: human factors in map design and interpretation. *Transactions of the Institute of British Geographers* 2(1): 19–36

Bowman A W (1973) A comparative study of some kernel-based nonparametric density estimators. *Journal of Statistical Computation and Simulation* 21: 313–327

Bowman A W (1984) An alternative method of cross-validation for the smoothing of density estimates. *Biometrika* 71: 353–360

Bracken I (1989) The generation of socioeconomic surfaces for public policy making. *Environment and Planning B* 16: 307–326

Bracken I (1991) A surface model of population for public resource allocation. *Mapping Awareness* 5(6): 35–39

Bracken I (1992) A surface model approach to the representation of population-related social indicators. Paper presented to Workshop I-14, San Diego. National Center for Geographic Information and Analysis, University of Buffalo, New York

Bracken I, Martin D (1989) The generation of spatial population distributions from census centroid data. *Environment and Planning A* 21: 537–543

Bracken I, Webster C (1990) *Information Technology in Geography and Planning*, Routledge, London

Brassel K (1974) A model for automatic hill shading. *The American Cartographer* 1(1): 15–27

Brewer C (1989) The development of process-printed Munsell charts for selecting map colors. *The American Cartographer* 16(4): 269–278

Brewer C A (1992) Review of Envisioning Information by Tufte E R. *Photogrammetric Engineering and Remote Sensing LVIII* (5): 544–545

Brodlie K W, Carpenter L A, Earnshaw R A, Gallop J R, Hubbold R J Mumford A M, Osland C D, Quarendon P (eds) (1992) *Scientific Visualization: Techniques and Applications*, Springer-Verlag, Berlin

Browne D, Totterdell P, Norman M (1990) *Adaptive User Interfaces*, Academic Press, London

Browne T J, Millington A C (1983) An evaluation of the use of grid squares in computerised choropleth maps. *The Cartographic Journal* 20: 71–75

Brunsdon C (1991) Estimating probability surfaces in GIS: an adaptive technique. In: Harts J, Ottens H F L, Scholten H J (eds) *Proceedings, Second European Conference on Geographical Information Systems*, EGIS Foundation, Utrecht, Netherlands, pp. 155–64

Bunge W (1966) *Theoretical Geography*. Lund Studies in Geography, Series C, 1: 285pp.

Burrough P A (1986) *Principles of Geographical Information Systems for Land Resources Assessment*, Oxford University Press, Oxford

Burrough P A, Heuvelink G B M (1992) The sensitivity of Boolean and continuous (fuzzy) logical modelling to uncertain data. *Proceedings, EGIS 92, Munich* 2: 1032–1041

Butler D M, Pendley M H (1989) A visualization model based on the mathematics of fiber bundles. *Computers in Physics* 3(5): 45–51

Buttenfield B P (1991) Visualizing cartographic metadata. In: Beard M K, Buttenfield B P, Clapham S (1991) *Visualizing the Quality of Spatial Information, Scientific Report of the Specialist Meeting*. NCGIA Technical Report 91–26, Santa Barbara, Calif.

Buttenfield B P, Beard M K (1991) Visualizing the quality of spatial information. *Proceedings AUTO-CARTO 10*, Baltimore Maryland 6: 423–427

Buttenfield B P, Ganter J H (1990) Visualization and GIS: What should we see? What might we miss? *Proceedings of the 4th International Symposium on Spatial Data Handling*, Zurich, Switzerland, pp. 307–316

Buttenfield B P, MacKaness W A (1992) Visualization. In: Maguire D J, Goodchild M F, Rhind D W (eds) *Geographical Information Systems: Principles and Applications,* pp. 19–36

Buttenfield B, McMaster R B (eds) (1991) *Map Generalization: Making Rules for Knowledge Representation,* Longman, Harlow

Campari I, Scopigno R, Magnarapa C (1990) Visual language approach: a proposal for improving user friendliness and effective use of GIS. *Proceedings of the First European GIS Conference,* Amsterdam, Netherlands, pp. 1024–1034

Campbell J B (1987) *Introduction to Remote Sensing,* Guildford, New York

Carroll J (1990) Task analysis: the oft missing step in the development of computer–human interfaces; its desirable nature, value and role, panel session report. In: Diaper D et al. (eds) *Human–Computer Interaction – Interact '90.* Elsevier, North-Holland, pp. 1051–1054

Carter J R (1989) Relative errors identified in USGS DEMs. *Proceedings, Auto Carto 9,* American Congress on Surveying and mapping, Bethesda, Maryland, 255–265

Castner H W (1979) Viewing time and experience as factors in map design research. *Canadian Cartographer* 16(2): 145–158

Census Research Unit (1980) *People in Britain – a census atlas,* HMSO, London

CERL (1991) *The GRASS 4.0 Reference Manual.* US Army Construction Engineering Research Laboratory, Champaign, Illinois

Chamberlin T C (1890) The method of multiple working hypotheses. *Science* 15: 92–96

Chernoff H (1973) The use of faces to represent points in k-dimensional space graphically. *Journal of the American Statistical Association* 68(342): 361–368

Chernoff H (1978) Graphical representations as a discipline. In: Wang P C C (ed.) *Graphical Representation of Multivariate Data,* Academic Press, New York

Chernoff H, Rizvi M H (1975) Effect on classification error of random permutations of features in representing multivariate data by faces. *Journal of the American Statistical Association* 70(351): 548–554

Chou Y-H (1991) Map resolution and autocorrelation. *Geographical Analysis* 23(3): 228–245

Chrisman N R (1983) The role of quality information in the long-term functioning of a geographic information system. *Cartographica* 21 (2–3): 79–87

Chrisman N R (1987) Fundamental principles of geographic information systems. In: *Proceedings Auto Carto 8,* American Society for Photogrammetry and Remote Sensing, Falls Church, Virginia, pp. 32–41

Chrisman N R (1989) Modeling error in overlaid categorical maps. In: Goodchild M F, Gopal S (eds) *The Accuracy of Spatial Databases,* pp. 21–34

Clapham S (1992) *A formal approach to the visualization of spatial data quality.* Unpublished MSc Thesis, Department of Surveying Engineering, University of Maine

Cleveland W S, McGill M E (1988) *Dynamic Graphics for Statistics,* Wadsworth & Brooks/Cole Advanced Books & Software, Belmont, Calif.

Cliff A D, Ord J K (1981) *Spatial Processes: Models and Applications,* Pion. London

Coppock J T, Anderson E K (1987) Editorial review. *International Journal of Geographical Information Systems* 1(1): 3–11

Coulson M R (1977) Political truth and the graphic image. *The Canadian Cartographer* 14(20): 101–111

Craig J (1976) 1971 census grid squares. *Population Trends* 6: 14–15

Cressie N A C (1991) *Statistics for Spatial Data*, Wiley, New York

Cressman G P (1959) An operational objective analysis system. *Monthly Weather Review* 87(10): 367–374

Cuff D J (1972) Value versus chroma in color schemes on quantitative maps. *Canadian Cartographer* 9(2): 134–140

Cuff D J (1973) Shading on choropleth maps: some suspicions confirmed. *Proceedings, Association of American Geographers*, Annual Meeting 24 (April): 50–54

Cuff D J (1974) Impending conflict in color guidelines for maps of statistical surfaces. *Canadian Cartographer* 11(1): 54–58

Cuff D J, Mattson M T (1982) *Thematic Maps: Their Design and Production*, Methuen, New York

Dawsey C B III (1990) Algorithms for uniform range interval classification. *Cartographica* 27(1): 46–53

Dent B D (1990) *Cartography, Thematic Map Design*, WCB Publishers, Dubuque

Department of the Environment (DoE) (1987) *Handling Geographic Information*, HMSO, London

Devereux B J (1986) The integration of cartographic data stored in raster and vector formats. In: *Proceedings, Auto Carto London*, Royal Institution of Chartered Surveyors, London, pp. 257–266

Diaper D (1989) *Task Analysis for Human Computer Interaction*, Ellis Horwood, Chichester

DiBiase D (1990) Visualization in the earth sciences. *Earth and Mineral Sciences*, Bulletin of the College of Earth and Mineral Sciences, Pennsylvania State University 59(2): 13–18

DiBiase D, Krygier J, Reeves C, MacEachren A, Brenner A (1991a) *An Elementary Approach to Cartographic Animation*. Video presented at the Annual Meeting of the Association of American Geographers, April 13–17.

DiBiase D, MacEachren A, Krygier J, Reeves C (1992) Animation and the role of map design in scientific visualization. *Cartography and Geographical Information Systems* 19(4): 201–214

DiBiase D, MacEachren A, Krygier J, Reeves C, Brenner A (1991b) Animated cartographic visualization in earth system science. *Proceedings of the 15th International Cartographic Association Conference*, Bournemouth, UK, pp. 223–232

Diggle P J (1985) A kernel method for smoothing point process data. *Applied Statistics* 34: 138–147

Diggle P J (1990) A point process modelling approach to raised incidence of a rare phenomenon in the vicinity of a pre-specified point. *Journal of the Royal Statistical Society, Series A* 153: 349–62

Diggle P J, Gatrell A C, Lovett A A (1990) Modelling the prevalence of cancer of the larynx in part of Lancashire: a new methodology for spatial epidemiology. In: Thomas R W (ed.) *Spatial Epidemiology*, Pion, London, pp. 35–47

Dikau R (1989) The application of a digital relief model to landform analysis in geomorphology. In: Raper J (ed.) *Three-Dimensional Applications in GIS*, Taylor and Francis, London, pp. 51–77

Dobson M W (1973) Choropleth maps without class intervals? A comment. *Geographical Analysis* 5: 358–360

Dorling D (1990) A cartogram for visualization. *North East Regional Research Laboratory Research Reports* 90/4, University of Newcastle upon Tyne, Newcastle upon Tyne UK

Dorling D (1991) *The visualization of spatial social structure*. Unpublished PhD thesis, University of Newcastle upon Tyne, UK

Dorling D (1992) Stretching space and splicing time: from cartographic animation to interactive visualization. *Cartography and Geographical Information Systems* 19(4): 215–227

Dorling D, Openshaw S (1991) Experiments using computer animation to visualize space–time patterns. In: Klosterman R E (ed.) *Proceedings of Second International Conference on Computers in Urban Planning and Management*, Oxford, pp. 391–406

Doslak W Jr, Crawford P V (1977) Color influence on the perception of spatial structure. *Canadian Cartographer* 14(2): 120–129

Dougenik J A, Chrisman N R, Niemeyer D R (1985) An algorithm to construct continuous area cartograms. *Professional Geographer* 37(1): 75–81

Draper S, Waite D (1991) Iconographer as a visual programming system. In: Diaper D, Hammond N (eds) *People and Computers VI*, Cambridge University Press, Cambridge UK, pp. 171–185

Dutton G (1989) Modelling locational uncertainty via hierarchical tesselation. In: Goodchild M F, Gopal S (eds) *Accuracy of Spatial Databases*, pp. 125–140

Dutton G (1992) Handling positional uncertainty in spatial databases. *Proceedings, 5th Spatial Data Handling Symposium*, Charleston, South Carolina, 2: 460–469

Dykes J A (1991) *An Investigation Into the Effects of Classification Upon Spatial Autocorrelation in an Attempt to Maximise the 'Communication Effectiveness' of the Choropleth Map*. Unpublished MSc Thesis, University of Leicester UK

Earnshaw R A, Wiseman N (eds) (1992) *An Introductory Guide to Scientific Visualization*, Springer-Verlag, Berlin

Eastman J R, Nelson W, Shields G (1981) Production considerations in isodensity mapping. *Cartographica* 18(1), 24–30

Edwards B (1987) *Drawing on the Artist Within*, Collins, London

Edwards B (1989) *Drawing on the Right Side of the Brain*, Tarcher, Los Angeles

Edwards G (1992) Visualization – the second generation, *Image Processing* 24 (May/June): 48–53

Egbert S C, Slocum T A (1992) EXPLOREMAP: An exploration system for choropleth maps. *Annals, Association of American Geographers* 82(2): 275–288

Englund E (1993) Spatial simulation: environmental applications. In: Goodchild M F, Parks B O, Steyaert L T (eds) *Environmental Modeling with GIS*, Oxford University Press, New York

Epstein E, Roitman H (1987) Liability for information. *Proceedings Annual Meetings of the Urban and Regional Information Systems Association (URISA)* pp. 115–125

Evans G W, Pezdek K (1980) Cognitive mapping: knowledge of mapping and real world distance and location information. *Journal of Experimental Psychology, Human Learning and Memory* 6: 3–24

Evans I S (1977) The selection of class intervals. *Transactions, Institute of British Geographers* 2(1): 98–124

Evans I S (1979) An integrated system of terrain analysis and slope mapping. *Final report on grant DA-ERO-591-73-G0040*, University of Durham UK

Evans I S (1980) An integrated system of terrain analysis and slope mapping. *Zeitschrift fur Geomorphologie* Suppl-Bd 36: 274–295

Ferguson E S (1977) The mind's eye: nonverbal thought in technology. *Science* 197(4306): 827–836

Fisher P F (1989) Knowledge-based approaches to determining and correcting areas of unreliability in geographic databases. In: Goodchild M F, Gopal S (eds) *The Accuracy of Spatial Databases*, pp. 45–54

Fisher P F (1991a) First experiments in viewshed uncertainty: the accuracy of the viewshed area. *Photogrammetric Engineering and Remote Sensing* 57(10): 1321–1327

Fisher P F (1991b) Modelling soil map-unit inclusions by Monte Carlo simulation. *International Journal of Geographical Information Systems* 5(2): 193–208

Fisher P F (1991c) Spatial data sources and data problems. In: Maguire D J, Goodchild M F, Rhind D W (eds) *Geographical Information Systems: Principles and Applications*, pp. 175–189

Fisher P F (1992) First experiments in viewshed uncertainty: simulating fuzzy viewsheds. *Photogrammetric Engineering and Remote Sensing* 58(4): 345–352

Flowerdew R, Green M (1989) Statistical methods for inference between incompatible zonal systems. In: Goodchild M F, Gopal S (eds) *Accuracy of Spatial Databases*, Taylor and Francis, London, pp. 239–248

Flury B, Riedwyl H (1981) Graphical representation of multivariate data by means of asymmetrical faces. *Journal of the American Statistical Association* 76(376): 757–765

Flynn G W (1990) Chemical cartography: finding the keys to the kinetic labyrinth. *Science* 246: 1009–1015

Forbes J (1984) Problems of cartographic representation of patterns of population change. *The Cartographic Journal* 21(2): 93–102

Fotheringham A S, Wong D W S (1991) The modifiable areal unit problem in multivariate statistical analysis. *Environment and Planning A*, 23(7): 1025–1044

Frank A U (1991) Properties of geographic data: requirements for spatial access methods. In: Gunther O, Schek H-J (eds) *Advances in Spatial Databases: Proceedings of the 2nd Spatial Data Handling Symposium*, Zurich, pp. 225–234

Frisby J P (1979) *Seeing. Illusion, Brain and Mind*, Oxford University Press, Oxford

Ganter J H (1988) Interactive graphics: linking the human to the model. *Proceedings of GIS/LIS '88*, San Antonio, pp. 230–239

Gatrell A C (1974) *Complexity and Redundancy in Choropleth Maps.* Unpublished MSc thesis, Pennsylvania State University USA

Gatrell A C (1989) On the spatial representation and accuracy of address-based data in the United Kingdom. *International Journal of Geographical Information Systems* 3: 335–348

Gatrell A C (1991) Concepts of space and geographical data. In: Maguire D J, Goodchild M F, Rhind D W (eds) *Geographical Information Systems: Principles and Applications*, Longman, Harlow, pp. 119–34

Gatrell A C, Dunn C E, Boyle P J (1991) The relative utility of the Central Postcode Directory and Pinpoint Address Code in applications of geographical information systems. *Environment and Planning A* 23: 1447–1458

Gersmehl, P (1990) Choosing tools: nine metaphors for map animation. *Cartographic Perspectives* 5: 3–17

Gilbert G K (1896) The origin of hypotheses, illustrated by the discussion of a topographical problem. *Science* 3: 1–13

Gilmartin P P (1982) The instructional efficacy of maps in geographic text. *Journal of Geography* 4: 145–150

Goodchild M F (1985) Geographical information systems in undergraduate geography: a contemporary dilemma. *The Operational Geographer* 8: 34–38

Goodchild M F (1986) *Spatial Autocorrelation*, Concepts and Techniques in Modern Geography 47, Geo Books, Norwich UK

Goodchild M F (1989) Modelling error in objects and fields. In: Goodchild M F, Gopal S (eds) *Accuracy of Spatial Databases*, pp. 107–114

Goodchild M F (1991a) Integrating GIS and environmental modelling at global scales. *Proceedings of GIS/LIS '91*, Atlanta, Georgia, pp. 117–127

Goodchild M F (1991b) Issues of quality and uncertainty. In: Muller J-C (ed.) *Advances in Cartography*, Elsevier, New York, pp. 113–40

Goodchild M F, Gopal S (1989) (eds) *The Accuracy of Spatial Databases*, Taylor and Francis, London

Goodchild M F, Sun G, Yang S (1992) Development and test of an error model for categorical data. *International Journal of Geographical Information Systems* 6(2): 87–104

Gorny P, Tauber M J (1990) (eds) *Visualization in Human Computer Interaction: 7th Interdisciplinary Workshop on Informatics and Psychology*, Springer-Verlag, New York

Gould J D, Lewis C (1985) Designing for usability: key principles and what designers think. *Communications of the ACM* 28(3): 300–311

Gould M (1993) Two views of the user interface. In: Hearnshaw H, Medyckyj-Scott D J (eds) *Human Factors in Geographical Information Systems*, Belhaven Press, London

Gould M, McGranaghan M (1990) Metaphor in geographic information systems. *Proceedings Fourth International Symposium on Spatial Data Handling*, pp. 433–442

Gould P, DiBiase D, Kabel J (1990) Le SIDA: la carte animee comme rhetorique cartographique appliquee. *Mappe Monde* 90(1): 21–26

Green D G et al. (1989) A generic approach to landscape modelling. *Proceedings Simulation Society of Australia Conference*, Canberra

Haber R B, Lucas B, Collins N (1991) A Data Model for Scientific Visualization with Provisions for Regular and Irregular Grids, *Visualization 91 Proceedings*, IEEE Computer Society Press, Washington pp. 298–305

Haber R B, McNabb D A (1990) Visualization idioms: a conceptual model for scientific visualization systems. In: Nielson G M, Shriver B, Rosenblum L J (eds) *Visualization in Scientific Computing*, pp. 74–93

Haber R N, Wilkinson L (1982) Perceptual components of computer displays. *IEEE Computer Graphics and Applications*, May, pp. 23–35

Haggett P, Cliff A D, Frey A E (1977) *Locational Methods in Human Geography* 2nd edn, Edward Arnold, London

Haining R, Griffith D A, Bennett R (1983) Simulating two-dimensional autocorrelated surfaces. *Geographical Analysis* 15(3): 247–253

Hall S S (1992) *Mapping the Next Millenium: The Discovery of New Geographies*, Random House, New York

Hamming R W (1962) *Numerical Methods for Scientists and Engineers*. McGraw-Hill, New York

Haralick R M (1979) Statistical and structural approaches to texture. *Proceedings of the IEEE* 67(5): 786–804

Haralick R M, Shanmugam K, Dinstein I (1973) Textural features for image classification. *IEEE Transactions on Systems, Man and Cybernetics* SMC-3(6): 610–621

Harley J B (1990) Cartography, ethics and social theory. *Cartographica* 27(2): 1–27

Harrison M D, Thimbleby H W (eds) (1990) *Formal methods in HCI*, Cambridge University Press, Cambridge

Hayashibe K (1991) Reversals of visual depth caused by motion parallax. *Perception* 20(1): 17–28

Hayhurst E N, Milliron E L, Steiger J R (1977) *Soil Survey of Medina County, Ohio*, USDA-SCS, Washington DC

Hearnshaw H M (1991) Mental models of spatial databases. *Research Report* 27, Midlands Regional Research Laboratory, Leicester, UK

Helander M (1988) (ed.) *Handbook of Human–Computer Interaction*, Elsevier-North Holland, Amsterdam

Held R, Richards W (1972) *Perception: Mechanisms and Models*, Readings from *Scientific American*, Freeman & Co., San Francisco

Hodgson M E, Plews R W (1989) N-dimensional display of cluster means in feature space. *Photogrammetric Engineering and Remote Sensing* 55(5): 613–619

Hollingsworth T H (1964) The political colour of Britain by numbers of voters. *The Times*, 19 October, p. 18

Hollingsworth T H (1966) The political colour of Britain by winning parties. *The Times*, 4 April, p. 8

Hopgood F R A, Duce D A, Fielding E V C, Robinson K, Williams A S (1986) *Methodology of Window Management*, Springer-Verlag, Berlin

Iisaka J, Hegedus E (1982) Population estimation from Landsat imagery. *Remote Sensing of Environment* 12: 259–272

Imhof E (1982) *Cartographic Relief Representation*, Walter de Gruyter, Berlin

Jenks G F, Caspall F C (1971) Error on choroplethic maps: definition, measurement and reduction. *Annals, Association of American Geographers* 61: 217–244

Jenks G F, Knos D S (1961) The use of shading patterns in graded series. *Annals, Association of American Geographers* 51(3): 316–334

John E R, Prichep L S, Fridman J, Easton P (1988) Neurometrics: computer-assisted differential diagnosis of brain dysfunctions. *Science* 239: 162–169

Kabel J, Heidl R (1992) *AIDS in the United States, Past, Present, and Future, 1980–1995*, unpublished video

Kaneda K, Kato F, Nakame E, Nishita T, Tanaka H, Noguchi T (1989) Three-dimensional terrain modeling and display for environmental assessment. *Proceedings ACM Siggraph '89*, Boston

Keates J S (1982) *Understanding Maps*, Longman, Harlow

Keates J S (1989) *Cartographic Design and Production* 2nd edn, Longman, Harlow

Kellogg W A (1990) Qualitative artifact analysis. In: Diaper D et al. (eds) *Human–Computer Interaction – Interact '90* Elsevier, North-Holland, pp. 193–198

Kennedy S, Tobler W R (1983) Geographic interpolation. *Geographical Analysis* 15(2): 151–156

Kidner D B, Jones C B, Knight D G, Smith D H (1990) Digital terrain models for radio path profiles. *Proceedings of the 4th International Symposium on Spatial Data Handling*, Zurich, p. 240

Kolberg D W (1970) Population aggregations as a continuous surface. *Cartographic Journal* 7: 95–100

Kraak M J (1991) The cartographic functionality of a three-dimensional GIS. *Proceedings of the 15th ICA Conference*, Bournemouth, UK, pp. 917–921

Krygier J (1991) Sound variables, sound maps, and cartographic visualization. (Draft manuscript available via electronic mail from JBK5@PSUVM.PSU.EDU)

Lancaster P, Salkauskas K (1986) *Curve and Surface Fitting – An Introduction*, Academic Press, London

Langford M, Unwin D J, Maguire D J (1990a) Generating improved population density maps in an integrated GIS. *Proceedings of the First European Conference on Geographical Information Systems*, EGIS Foundation, Utrecht

Langford M, Unwin D J, Maguire D J (1990b) Mapping the density of population: continuous surface representations as an alternative to choroplethic and dasymetric maps. *Research Report* 8, Midlands Regional Research Laboratory, Leicester UK

Lanter D P, Veregin H (1990) A lineage meta-database program for propagating error in geographic information systems. *Proceedings GIS/LIS 90*, Anaheim, Calif. 1: 144–153

Laurini R, Thompson D (1992) *Fundamentals of Spatial Information Systems*, Academic Press, San Diego, Calif.

Lee J, Snyder P K, Fisher P F (1992) Modelling the effect of data errors on feature extraction from digital elevation models. *Photogrammetric Engineering and Remote Sensing* 58(10): 1461–1467

Leung Y (1988) *Spatial Analysis and Planning under Imprecision*, Elsevier, New York

Leung Y, Goodchild M F, Lin Chih-Chang (1992) Visualization of fuzzy scenes and probability fields. *Proceedings, Fifth International Symposium on Spatial Data Handling, Charleston South Carolina*, International Geographical Union, Columbus, Ohio

Levison M E, Haddon W (1965) The area adjusted map: an epidemiological device. *Public Health Reports* 80(1): 55–59

Liebenberg E (1976) SYMAP: its uses and abuses. *The Cartographic Journal* 13(1): 36–46

Lloyd R, Steinke T (1976) The decision making process for judging the similarity of choropleth maps. *The American Cartographer* 3(2): 177–184

Lo P (1989) A raster approach to population estimation using high-altitude aerial and space photographs. *Remote Sensing of Environment* 27: 59–71

MacDougall E B (1991) Dynamic statistical visualization of geographic information systems. *Proceedings of GIS/LIS '91*, Atlanta, Georgia, pp. 158–165

MacDougall E B (1992) Exploratory analysis, dynamic statistical visualization, and geographic information systems. *Cartography and Geographical Information Systems*, 19(4): 237–246

MacEachren A M (1992) Visualizing uncertain information. *Cartographic Perspectives* 13: 10–18

MacEachren A M, Buttenfield B, Campbell J, DiBiase D, Monmonier M (1992) Visualization. In: Abler R, Marcus M and Olson J (eds) *Geography's Inner Worlds: Pervasive Themes in Contemporary American Geography*, Rutgers University Press, New Brunswick, New Jersey, pp. 99–137

MacEachren A M, Davidson J V (1987) Sampling and isometric mapping of continuous geographic surfaces. *The American Cartographer* 14(4): 299–320

MacEachren A M, Ganter J H (1990) A pattern identification approach to cartographic visualization. *Cartographica* 27(2): 64–81

Mackay J R (1951) Some problems and techniques in isopleth mapping. *Economic Geography* 27(1): 1–9

Mackinlay J (1986) Automating the design of graphical presentations of relational information. *ACM Transactions on Graphics* 5: 110–141

Maguire D J (1991) An overview and definition of GIS. In: Maguire D J, Goodchild M F, Rhind D W (eds) *Geographical Information Systems: Principles and Applications*, pp. 2–20

Maguire D J, Goodchild M F, Rhind D W (eds) (1991) *Geographical Information Systems: Principles and Applications*, Longman, Harlow

Mark D M, Csillag F (1989) The nature of boundaries in 'area-class' maps. *Cartographica* 26(1): 65–78

Marsh S (1990) Human–computer interaction: an operational definition. *SIGCHI Bulletin* 22(1): 16–22

Marshall R, Kempf J, Dyer S, Yen Chieh-Cheng (1990) Visualization methods and simulation steering for a 3D turbulence model of Lake Erie. *Computer Graphics* 24(2): 89–97

Martin D (1988) An approach to surface generation from centroid-type data. *Technical Reports in Geo-Information Systems, Computing and Cartography* 5, Wales and South West Regional Research Laboratory, Cardiff UK

Martin D (1989) Mapping population data from zone centroid locations. *Transactions, Institute of British Geographers* 14(1): 90–97

Martin D, Bracken I (1991) Techniques for modelling population-related raster databases. *Environment and Planning A* 23: 1069–1075

Masser I (1990) The Regional Research Laboratory initiative: an update. In: *The Association for Geographic Information Yearbook*, Taylor and Francis, London, pp. 259–63

McCormick B H, DeFanti T A, Brown M D (eds) (1987) Visualization in scientific computing. Special issue ACM SIGGRAPH *Computer Graphics* 21(6)

McKim R H (1972) *Experiences in Visual Thinking*, Brooks/Cole, Monterey

McLaren R A, Kennie T J M (1989) Visualization of digital terrain models: techniques and applications. In: Raper J (ed.) *Three-Dimensional Applications in GIS*, Taylor and Francis, London, pp. 80–98

Medyckyj-Scott D J, Blades M (1991) Cognitive representations of space in the design and use of geographical information systems. In: Diaper D, Hammond N (eds) *People and Computers VI*, Cambridge University Press, Cambridge UK, pp. 421–433

Medyckyj-Scott D J, Walker D, Newman I, Ruggles C (1990) Usability testing in GIS and the role of rapid prototyping. *Proceedings of the First European GIS Conference*, Amsterdam, Netherlands, pp. 737–746

Moellering H (1973a) The automated mapping of traffic crashes. *Survey and Mapping* 33: 467–477

Moellering H (1973b) Computer animated film on traffic crashes Washtenaw County Michigan 1968–70. University of Michigan, Ann Arbor, Michigan

Moellering H (1976) The potential uses of a computer animated film in the analysis of geographical patterns of traffic crashes. *Accident Analysis & Prevention* 8: 215–227

Moellering H (1978) *A demonstration of the real time display of three-dimensional cartographic objects.* Computer-generated videotape, Department of Geography, Ohio State University, Columbus Ohio

Moellering H (ed.) (1988) The proposed standard for digital cartographic data: Report of the Digital Cartographic Data Standards Task Force. *The American Cartographer* 15(1)

Moellering H (1989) A practical and efficient approach to the stereoscopic display and manipulation of cartographic objects. *AUTO-CARTO 9, Proceedings, Ninth International Symposium on Computer-Assisted Cartography*, Baltimore, Maryland, pp. 1–4

Moellering H (1991) Stereoscopic display and manipulation of larger 3-D cartographic objects. *Proceedings of the 15th ICA Conference*, Bournemouth, UK, pp. 122–129

Monmonier M S (1974) Measures of pattern complexity for choroplethic maps. *The American Cartographer* 1(2): 159–169

Monmonier M (1989a) Geographic brushing: enhancing exploratory analysis of the scatterplot matrix. *Geographical Analysis* 21: 81–84

Monmonier M (1989b) Graphic scripts for the sequenced visualization of geographic data. *Proceedings of GIS/LIS '89*, Orlando, Florida, pp. 381–389

Monmonier M (1990) Strategies for the interactive exploration of geographic correlation. *Proceedings of the 4th International Symposium on Spatial Data Handling*, Zurich, Switzerland, pp. 512–521

Monmonier M S (1991) *How to Lie with Maps*, University of Chicago Press, Chicago Illinois

Monmonier M (1992) Authoring graphic scripts: experiences and principles. *Cartography and Geographical Information Systems*, 19(4): 247–260

Monmonier M, MacEachren A M (eds) (1992) Special content: Geographic Visualization. *Cartography and Geographic Information Systems* 19(4): 197–260

Moore T J (1991) Application of GIS technology to air toxic risk assessment: meeting the demands of the California Air Toxics 'Hot Spot' Act of 1987. *Proceedings of GIS/LIS '91*, Atlanta, Georgia, pp. 694–714

Muehrcke P C (1972) *Thematic Cartography*. Commission on College Geography, Resource Paper 19, Association of American Geographers, Washington DC

Muehrcke P C (1973) The influence of spatial autocorrelation and cross correlation on visual map comparison. *Proceedings of the American Congress on Mapping and Surveying*, pp. 315–325

Muehrcke P C (1978) *Map Use: Reading, Analysis and Interpretation*, J P Publications, Madison Wisconsin

Muehrcke P (1981) Maps in geography. *Cartographica* 18(2): 1–41

Muller J-C (1975) Associations in choropleth map comparison. *Annals, Association of American Geographers* 65(3): 403–415

Muller J-C (1976a) Number of classes and choropleth pattern characteristics. *The American Cartographer* 3(2): 169–175

Muller J-C (1976b) Visual association in choropleth mapping. *Proceedings of the Association of American Cartographers* 8: 160–164

Muller J-C (1992) Generalization of spatial databases. In: Maguire D J, Goodchild M F, Rhind D W (eds) *Geographical Information Systems: Principles and Applications*, pp. 457–475

Neilson G M, Shriver B, Rosenblum L (eds) (1990) *Visualization in Scientific Computing*, IEEE Computer Society Press, Los Alamitos, Calif., Washington

Nordbeck S, Rystedt B (1970) Isarithmic maps and the continuity of reference interval functions. *Geografiska Annaler* 52B: 92–123

Norman D A (1991) Cognitive artifacts. In: Carroll J (ed.) *Designing Interaction: Psychology at the Human–Computer Interface*, Cambridge Series on Human–Computer Interaction 4, Cambridge University Press, Cambridge UK, pp. 17–38

Norman D A, Draper S W (1986) (eds) *User-centred System Design: New Perspectives on Human Computer Interaction*, Lawrence Erlbaum Associates, Hillsdale NJ

Norman K L, Weldon L J, Shneiderman B (1986) Cognitive layout of windows and multiple screens for user interfaces. *International Journal of Man–Machine Studies* 25(2): 229–248

Oberg S, Springfeldt P (eds) (1991) *The Population Atlas of Sweden*, SNA, Stockholm

Oden N L (1984) Assessing the significance of a spatial correlogram. *Geographical Analysis* 16(1) 1–16

Olson J M (1974) Autocorrelation and visual map complexity. *Annals, Association of American Geographers* 65(2): 189–204

Olson J M (1976) Noncontiguous area cartograms. *Professional Geographer* 28: 371–380

Olson J R, Olson G M (1990) The growth of cognitive modelling in human–computer interaction since GOMS. *Human–Computer Interaction* 5: 221–265

Openshaw S (1984) *The Modifiable Areal Unit Problem*. Concepts and Techniques in Modern Geography 38, Geo Books, Norwich UK

Openshaw S (1992) Developing appropriate spatial analytical methods for GIS. In: Maguire D J, Goodchild M F, Rhind D W (eds) *Geographical Information Systems: Principles and Applications*, pp. 389–402

Ordnance Survey (1992) *1:50 000 Scale Height Data User Manual*, Ordnance Survey, Southampton UK

Orland B, Onstad D, Obermark J, LaFontaine J (1992) Visualization of plant growth and pest models. *Technical Papers*, ASPRS/ACSM/RT 92, 5, Resource Technology, 246–253

Ottoson L, Rystedt B (1980) Computer-assisted cartography: research and applications in Sweden. In: Taylor D R F (ed.) *The Computer in Contemporary Cartography*, Wiley, Chichester, pp. 93–122

Palmer T C (1992) A language for molecular visualization. *IEEE Computer Graphics and Applications* 12(3): 23–32

Parker N (1993) GIS hardware design. In: Hearnshaw H, Medyckyj-Scott D J (eds) *Human Factors in Geographical Information Systems*, Belhaven Press, London

Peterson M P (1985) Evaluating a map's image. *The American Cartographer* 12(1): 41–55

Peucker T K, Fowler R J, Little J J, Mark D M (1978) The triangulated irregular network. *Proceedings of the DTM Symposium*, American Congress on Surveying and Mapping, St Louis, Missouri, pp. 24–31

Peuquet D J (1988) Representations of geographic space: toward a conceptual synthesis. *Annals, Association of American Geographers* 78: 375–394

Phillips R J, De Lucia A, Skelton N (1975) Some objective tests of the legibility of relief maps. *The Cartographic Journal* 12(1): 39–46

Piwowar J M, LeDrew E F, Dudycha D J (1990) Integration of spatial data in vector and raster formats in geographical information systems. *International Journal of Geographical Information Systems* 4(4): 429–444

Platt J R (1964) Strong inference. *Science* 146: 347–353

Popper K R (1974) *Conjectures and Refutations*, 5th edn, Routledge and Kegan Paul, London

Popper K R (1975) *Objective Knowledge*, Oxford University Press, Oxford

Popper K R, Eccles J (1977) *The Self and its Brain*, Hutchinson, London

Porter P, Voxland P (1986) Distortion in maps: the Peters projection and other devilments. *Focus* 36: 22–30

Press W H, Flannery B P, Teukolsky S A, Vetterling W T (1988) *Numerical Recipes In C – The Art of Scientific Computing*, Cambridge University Press, Cambridge UK

Raisz E (1934) The rectangular statistical cartogram. *Geographical Review* 24: 292–296

Raisz E (1936) Rectangular statistical cartogram of the world. *Journal of Geography* 35: 8–10.

Rase W D (1987) The evolution of a graduated symbol software package in a changing graphics environment. *International Journal of Geographical Information Systems* 1: 51–66

Rasmussen J, Pejtersen A M, Schmidt K (1990) *Taxonomy for Cognitive Work Analysis*, Cognitive Systems Group, Riso National Laboratory, Roskilde, Denmark

Reed S K (1973) *Psychological Processes in Pattern Recognition*, Academic Press, London

Reeves W T, Blau R (1985) Approximate and probabilistic algorithms for shading and rendering structured particle systems. *Proceedings of SIGGRAPH' 85*, pp. 313–322

Rhind D (1983) Mapping census data. In: Rhind D (ed.) *A Census User's Handbook*, Methuen, London

Robbins R G, Thake J E (1988) Coming to terms with the horrors of automation. *The Cartographic Journal* 25: 139–142

Robertson A, Sale R, Morrison J (1978) *Elements of Cartography*, Wiley, New York

Robertson P K (1988a) Choosing data representations for the effective visualization of spatial data. *Proceedings, Third International Symposium on Spatial Data Handling*, Sydney, Australia, pp. 243–252

Robertson P K (1988b) Visualizing color gamuts: A user interface for the effective use of perceptual color spaces in data displays. *IEEE Computer Graphics and Applications*, September, pp. 50–64

Robertson P K (1990) A methodology for scientific data visualization: choosing representations based on a natural scene paradigm. In: *Proceedings of the 1st IEEE Conference on Visualization – Visualization '90*, San Francisco, IEEE Computer Press, Washington, pp. 114–123

Robertson P K (1991) A methodology for choosing data representations. *IEEE Computer Graphics and Applications* 11(3): 56–67

Robinson A H (1952) *The Look of Maps*, University of Wisconsin Press, Madison

Robinson A H (1967) Psychological aspects of colour in cartography. *International Yearbook of Cartography* VIII: 50–61

Robinson A H (1982) *Early Thematic Mapping in the History of Cartography*, University of Chicago Press, London

Robinson A H, Petchenik B B (1975) The map as a communication system. *The Cartographic Journal* 12: 7–14

Robinson A H, Petchenik B B (1976) *The Nature of Maps*, University of Chicago Press, London

Robinson A H, Sale R, Morrison J, Muehrcke P (1985) *Elements of Cartography*, 5th edn, Wiley, Chichester

Rosenblum L J, Nielson G M (1991) Guest editor's introduction: visualization comes of age. *IEEE Computer Graphics and Applications* 11(3): 15–17

Rowlingson B S, Diggle P J (1991) SPLANCS: spatial point pattern analysis code in S-Plus. *North West Regional Research Laboratory, Research Report 22*, Lancaster University, Lancaster UK

Rowlingson B S, Flowerdew R, Gatrell A C (1991) Statistical spatial analysis in a Geographical Information Systems framework. *North West Regional Research Laboratory, Research Report 23*, Lancaster University, Lancaster UK

Samet H (1984a) Algorithms for the conversion of quadtrees to raster. *Computer Graphics and Image Processing* 26: 1–16

Samet H (1984b) The quadtree and related hierarchical data structures. *ACM Computing Survey* 16: 187–260

Sarjakoski T (1990) Digital stereo imagery – integration to geo-information systems. *Proceedings of Commission 3, XVI FIG Congress*, Helsinki, Finland, pp. 489–498

Schalkoff R J (1989) *Digital Image Processing and Computer Vision*, Wiley, New York

Schmid C F, MacCannell E H (1955) Basic problems, techniques and theory in isopleth mapping. *Journal of the American Statistical Association* 50: 220–239

SDTS (1990) *The Spatial Data Transfer Standard.* US Government Printing Office, US Geological Survey Bulletin, Washington DC

Sen A K (1975) A theorem related to cartograms. *American Mathematics Monthly* 82: 382–385

Shackel B (1983) The concept of usability. In Bennett J et al. (eds) *Visual Display Terminals: Usability Concerns and Health Issues*, Prentice-Hall, New Jersey, pp. 45–87

Shepard R N (1978) Externalisation of mental images and the act of creation. In: Randhawa B S, Coffman W E (eds) *Visual Learning, Thinking and Communication*, Academic Press, London

Shirayev E E (1987) *Computers and the Representation of Geographical Data*, Wiley, Chichester

Silverman B W (1986) *Density Estimation for Statistics and Data Analysis*, Chapman and Hall, London

Silverstone (1991) Interview with Tufte. *Aldus Magazine*, May/June. Cited by Brewer C A (1992) Review of Envisioning Information by Tufte E R. *Photogrammetric Engineering and Remote Sensing* LVIII: 544–545

Sinton D F (1977) *The Users Guide to IMGRID. An Information Manipulation System for Grid Cell Structures*, Harvard University, Department of Landscape Architecture, Cambridge, Mass.

Sinton D (1978) The inherent structure of information as a constraint to analysis. In: Dutton D (ed.) *Harvard Papers on Geographic Information Systems*, Addison Wesley, Reading

Skidmore A K (1989) A comparison of techniques for calculating gradient and aspect from a gridded digital elevation model. *International Journal of Geographical Information Systems* 3(4): 323–334

Skoda L, Robertson J C (1972) Isodemographic map of Canada. *Geographical Papers 50*, Department of the Environment, Ottawa, Canada

Slocum T A, Robertson S H, Egbert S L (1990) Traditional versus sequenced choropleth maps: an experimental investigation. *Cartographica* 27(1): 67–88

Smart J, Mason M (1990) Assessing the visual impact of development plans. In: Heit M, Shortreid A (eds) *GIS applications in natural resources*, GIS World, Inc., Fort Collins, pp. 295–303

Stephenson T (1990) Imaging, visualization and the challenge of global change: this technology's role in the Mission to Planet Earth. *Advanced Imaging* 5(7): 59–61

Stevens S S (1959) Cross-modality validation of subjective scales for loudness, vibration and electric shock. *Journal of Experimental Psychology* 57: 201–209

Szego J (1987) *Human Cartography: Mapping the World of Man*, Swedish Council for Building Research, Stockholm

Teicholz E, Berry B J L (eds) (1983) *Computer Graphics and Environmental Planning*, Prentice-Hall, Englewood Cliffs NJ

ten Velden H E, van Lingen M (1990) Geographical information systems and visualization. In: Scholten H, Stillwell J C (eds) *GIS for Urban and Regional Planning*, Kluwer Press, Holland, pp. 229–237

Theobald D (1989) Accuracy and bias in surface representation. In: Goodchild M F, Gopal S (eds) *Accuracy of Spatial Databases*, pp. 99–106

Thompson M M (1988) *Maps for America*, 3rd edn, USGS Reston, Virginia

Thorndyke P W, Stasz C (1980) Individual differences in procedures for knowledge acquisition from maps. *Cognitive Psychology* 12: 137–175

Thrower N (1959) Animated cartography. *The Professional Geographer* 11(6): 9–12

Tobler W R (1961) *Map transformations of geographic space*. Unpublished PhD thesis, Department of Geography, University of Washington USA

Tobler W R (1963) Geographic area and map projections. *Geographical Review* 53: 59–78

Tobler W R (1970) A computer movie simulating urban growth in the Detroit region. *Economic Geography* 46: 234–240

Tobler W R (1973) Choropleth maps without class intervals? *Geographical Analysis* 5: 262–265

Tobler W R (1976) Cartograms and cartosplines. *Proceedings of the Workshop on Automated Cartography and Epidemiology*, National Center for Health Statistics, US Department of Health, pp. 53–58

Tobler W R (1979) Smooth pycnophylactic interpolation for geographic regions. *Journal of the American Statistical Association* 74: 519–530

Tobler W R (1981) Depicting federal fiscal transfers. *Professional Geographer* 33(4): 419–422

Tobler W R (1985) Interactive construction of continuous cartograms. *Proceedings of the 7th International Symposium on Computer-Assisted Cartography*, Washington DC, p. 525

Tobler W R (1986) Pseudo-cartograms. *The American Cartographer* 13(1): 43–50

Tobler W (1987) Measuring spatial resolution. *Proceedings of International Workshop on Geographical Information Systems*, Beijing, pp. 42–48

Tomlin C D (1991) *Geographic Information Systems and Cartographic Modeling*, Prentice-Hall, Englewood Cliffs, New Jersey

Travis D S (1990) Applying visual psychophysics to user interface design. *Behavior & Information Technology* 9(5): 425–438

Trotter C M (1991) Remotely-sensed data as an information source for geographical information systems in natural resource management: a review. *International Journal of Geographical Information Systems* 5(2): 225–239

Tufte E R (1983) *The Visual Display of Quantitative Information*, Graphic Press, Cheshire, Connecticut

Tufte E R (1990) *Envisioning Information*, Graphics Press, Cheshire, Connecticut

Turk A G (1990) Towards an understanding of human–computer interaction aspects of geographic information systems. *Cartography* 19(1): 31–60

Turk A G (1992) *GIS cogency: Cognitive ergonomics in geographic information systems*, Unpublished PhD thesis, University of Melbourne, Australia

Unwin D (1981) *Introductory Spatial Analysis*, Methuen, London

Upson C, Faulhaber T Jr, Kamins D, Laidlaw D, Schlegel D, Vroom J, Gurwitz R, van Dam A (1989) The application visualization system: a computational environment for scientific visualization. *IEEE Computer Graphics and Applications* 9(4): 30–42

Upton G J G (1991) Displaying election results. *Political Geography Quarterly* 10(3): 200–220

US Geological Survey (1987) *Digital Elevation Models – Data Users Guide*, US Geological Survey, Reston, Virginia

van Elzakker C P J M (1991) Map use research and computer-assisted statistical cartography. *Proceedings of the 15th ICA Conference*, Bournemouth, UK, pp. 575–584

Vasiliev I (1991) The dimensionality of time in cartographic thought. Paper presented at the Annual Meeting of the Association of American Geographers, Miami, Florida

Visvalingam M (1989) Cartography, GIS and maps in perspective. *Cartographic Journal* 26(1): 26–32

Visvalingam M (1990) Trends and concerns in digital cartography. *Computer-Aided Design* 22(3): 115–130

Visvalingam M, Kirby G H (1984) *The Impact of Advances in IT on the Cartographic Interface in Social Planning*, Department of Geography Miscellaneous Series 27, University of Hull, Hull UK

Visvalingam M, Whyatt J D (1991) Cartographic algorithms: problems of implementation and evaluation and the impact of digitising errors. *Computer Graphics Forum* 10(3): 225–235

Walsh S J, Lightfoot D R, Butler D R (1987) Recognition and assessment of error in geographic information systems, *Photogrammetric Engineering and Remote Sensing* 53: 1423–1430

Weibel R, Buttenfield B P (1992) Improvement of GIS graphics for analysis and decision-making. *International Journal of Geographical Information Systems* 6(3): 223–245

Wood M (1990) Map perception studies. In: Perkins C R, Parry R B (eds) *Information Sources in Cartography*, Bowker-Saur, London, pp. 441–452

Yeoli P (1967) Mechanisation in analytical hill-shading. *Cartographic Journal* 4: 82–88

Yeoli P (1986) Computer executed production of a regular grid of height points from digital contours. *The American Cartographer* 13(3): 219–229

Zeitlin M (1992) Visualization brings a new dimension to oil exploration and production. *Geobyte* 7(3): 36–39

Ziegler J, Bullinger H-J (1991) Formal models and techniques in human–computer interaction. In: Shackel B and Richardson S (eds) *Human factors for informatics usability*, Cambridge University Press, Cambridge, pp. 183–206

BIBLIOGRAPHY

Zube E H, Simcox D E, Law C S (1987) Perceptual landscape simulations history and prospect. *Landscape Journal* 6: 62–80

INDEX